SURVEYING CULTURES

SURVEYING CULTURES

Discovering Shared Conceptions and Sentiments

David R. Heise
Department of Sociology
Indiana University
Bloomington, Indiana

WILEY

A JOHN WILEY & SONS, INC., PUBLICATION

Published by John Wiley & Sons, Inc., Hoboken, New Jersey.
Published simultaneously in Canada.

Library of Congress Cataloging-in-Publication Data:

Heise, David R.
Surveying cultures : discovering shared conceptions and sentiments / David R. Heise.
p. cm.
Includes bibliographical references and index.
ISBN 978-0-470-47907-0
1. Social surveys. 2. Ethnology. I. Title.
HM538.H45 2010
306.072′3–dc22

2009031377

Printed in the United States of America

10 9 8 7 6 5 4 3 2 1

Dedicated to
culture surveyors, past, present, and future

Contents

Preface

Methodologies for surveys that focus on differences among individuals made giant gains in the last half century. Survey research establishments proliferated, both profit and nonprofit, with the number of the latter growing from a half dozen to a hundred or more (O'Rourke, Sudman, and Ryan 1996). Stabilization of large survey organizations fostered development of serial surveys for over-time comparisons: both trend studies such as the General Social Survey and multidecade panel studies (Weisberg 2005, p. 9). Experimental studies and conversational analysis of interviews improved item and questionnaire construction (see Weisberg 2005, Part 2). Incorporation of new communication technologies—telephones, the Internet, cellular phones—improved surveys' efficiency and created new types of studies (e.g., see Schonlau, Fricker, and Elliot 2002). Survey researchers mustered mathematical statistics to provide practical procedures for dealing with a variety of sampling issues. Absorption of the computer revolution decreased survey costs, allowed interviews to be more logical and simultaneously more complex, facilitated data-gathering tools such as graphic rating scales, and provided a large tool chest of powerful statistical and graphing software for analyzing survey data.

Denzin and Lincoln (2005) argue that a revolution also occurred in qualitative research, the branch of social science focusing on meanings and culture. However, the vogue in cultural anthropology and in the sociology of culture is to penetrate the dynamics of culture construction and reproduction, and the revolution in this area largely concerns philosophic framing and political positioning of such studies, expanded notions of data, and acceptable forms of exposition. Aside from audio and visual recording, and software for coding recordings and texts, technological advances are viewed primarily as contextual influences on culture production, thereby becoming topics of investigation more than tools empowering investigations. That no person is an adequate informant regarding culture is generally understood, but no implication has been drawn that culture must be studied by surveying multiple persons. The Denzin and Lincoln (2005) handbook contains no chapter on survey research, and the methods chapter in a handbook of symbolic interactionism (Herman-Kinney and Verschaeve 2003) provides no discussion of surveys.

Researchers focusing on culture construction and reproduction have leveraged essential critiques of traditional ethnographies in such a way as almost to eliminate studies of established cultures and social organizations: that is, the enduring mental and social products of culture production. "By romanticizing the emergent and the immediate, this neo-vitalist position tends too briskly to dismiss given social formations as always already foreclosed" (Mazzarella 2004, p. 348). However, as Kashima (2002, pp. 208–211) emphasized (resonating a conception of Goodenough 1961), two standpoints characterize studies of culture and meaning. One standpoint presents culture as continuously produced and reproduced by fluctuating and yet recurrent processes of meaning-making, conducted by concrete individuals in particular contexts. Another standpoint highlights the persistence of culture over time, focusing on an enduring system of meanings that organizes people's shared experiences.

This book focuses on methodology involved in the second focus of culture studies: enduring systems of meanings. Researching enduring systems of meaning is sometimes like survey research in depending on questioning of informants and administration of carefully crafted questionnaires to indigenes. However, over the last half century nothing like the seismic shifts in survey research methodology occurred in methodologies for recording and analyzing persisting cultural forms (except perhaps in lexicography; see Landau 2001). Some methodological advances were made in ethnoscience (Werner and Fenton 1970) and cognitive anthropology (D'Andrade 1995): for example, Romney, Weller, and Batchelder's (1986) insight regarding sampling of informants that launched the work that I report in this book. In sociology, procedures were developed for studying sentiment norms and norms involved in processing events, as reviewed in the first four chapters of this book. Yet considerable murkiness still pervades the methodology of studying enduring systems of meaning, partly because the appropriation of culture studies by investigators of construction and reproduction left few resources and little interest for developing the methodology of normative studies. I hope that this book contributes to remedying that state of affairs by inciting further work on methods of recording and analyzing cultural norms.

Wiley maintains an ftp site for this book so that related files can be downloaded via the World Wide Web. The address is ftp://ftp.wiley.com/public/sci_tech_med/surveying_cultures. I have posted the following materials at the site:

- Image files of *Atlas* tables for 17 cultures identified in Section 3.2, plus sentiment measurements from these cultures in three electronic formats: SPSS, Stata, and comma-delimited text.
- The Java code for the Surveyor data-collection program discussed in Section 2.3, along with a Perl program called "transform" for receiving and archiving data via the Internet; some documentation also is included.
- FileMaker files containing male and female sentiment means and standard deviations for concepts rated in U.S.A., Canada, Ireland, Germany, Japan, and China.

June 2009 DAVID R. HEISE

Acknowledgments

I am deeply grateful to those who contributed to the data-collection effort that provided empirical grounding for Chapters 6 though 8. Clare Francis and Professor P. Christopher Earley recruited more than half of the respondents for the empirical study from Professor Earley's management class in the Kelley School of Business at Indiana University. (Francis is now a professor in the University of North Dakota's College of Business and Public Administration.) Professor Martin Laubach recruited additional respondents from his sociology classes at the University of Connecticut, providing valuable diversification in the set of respondents.

Professor Herman Smith of the Department of Sociology at the University of Missouri at St. Louis read an early draft of the manuscript and gave me sound advice as well as heartfelt encouragement to continue the project. Professor Adam King, Department of Sociology, Northern Illinois University, read a later draft and provided valuable suggestions that improved the content and exposition of the arguments. Professor Duane Alwin of Pennsylvania State University read the penultimate draft of the manuscript and offered important suggestions on how to improve the scope of the disquisition. Professor William Batchelder of the University of California–Irvine provided me with useful information regarding the culture-as-consensus paradigm. I am indebted to all these people for their generous professional help.

Additionally I gratefully acknowledge permissions from the copyright holders to quote selections from previously published texts, as indicated below.

Charles E. Osgood, W. H. May, and M. S. Miron, *Cross-Cultural Universals of Affective Meaning*. Copyright © 1975 by the Board of Trustees of the University of Illinois Press. Used with permission of the University of Illinois Press.

Wolfgang Scholl, *The Socio-emotional Basis of Human Cognition, Communication, and Interaction*. Berlin: Humboldt-University, Institute of Psychology, 2008. Reprinted by permission of Wolfgang Scholl.

David Heise, The semantic differential and attitude research. Pp. 235–253 in *Attitude Measurement*, edited by Gene Summers. Chicago: Rand McNally, 1970. Reprinted by permission of Gene F. Summers.

David Heise, "Face synthesizer," *Micro: The 6502/6809 Journal* 49:31–37, Copyright © June 1982 by FlexAble Systems, Inc., Fountain Hills, AZ, FKA The Computerist, Inc., DBA MICRO INK; reprinted by permission of Robert M. Tripp.

Peter Rossi and Andy Anderson, "The factorial survey approach: an introduction," pp. 15–67 in *Measuring Social Judgments: The Factorial Survey Approach*, edited by Peter H. Rossi and Steven L. Nock. Beverly Hills, CA: Sage Publications, 1982. Reprinted by permission of the publisher.

1 Surveying Culture

In traditional sample surveys, one person's measurements are presumed unpredictable from measurements on another person. For example, one person's age offers no clues about another person's age. Consequently, knowledge about variations and central tendencies in a population has to be built up piecemeal from many individual measurements. A statistical methodology has developed to guide the process of making inferences about a population from a sample of people (Kish 1965), and this methodology is so fundamental in social science that it frequently is treated as the only viable framework for acquiring and interpreting survey data.

However, work in psychological anthropology (Romney 1994; Romney, Batchelder, and Weller 1987; Romney, Weller, and Batchelder 1986) and in sociology (Rossi and Nock 1982) has clarified that some surveys are conducted in order to ascertain normative features shared by everyone, such as beliefs and sentiments deriving from culture. In this case, information from one person does predict information from others: For example, one person in a traditional society reporting that fathers usually are husbands foreshadows others saying the same thing. When people all provide the same information, it is redundant to ask a question over and over. Only enough people need be surveyed to eliminate the possibility of errors and to allow for those who might diverge from the norm. Romney, Weller, and Batchelder's (1986) mathematical-statistical analysis of ethnographic data gathering demonstrated that as few as a half-dozen expert respondents can provide a very clear picture of some types of shared norms.

In a survey of a population of individual subjects, variability is sought in the answers to every item, and items that would yield meager variability generally are avoided because they seem uninteresting. However, surveys that have been engineered to maximize variability fail to reveal norms.

> The measurement practices that have dominated the fields of sociology and social psychology seem designed to avoid finding empirical evidence for norms, beliefs about which there is some large degree of popular consensus. Indeed, when we find measures on which individual subjects agree, we tend to discard them because they will reveal little about differences among individuals. . . . We

Surveying Cultures: Discovering Shared Conceptions and Sentiments, By David R. Heise

> prefer items with maximum variance and hence with corresponding minimum agreement among subjects, a strategy that makes good sense in measuring inter-individual variation in the amount of cognitive achievement, but may not make good sense for sociologists who are trying to understand the overall normative patterning of human behavior. Indeed, it is often precisely those measurement instruments that we conventionally reject as useless that are most indicative of norms. Thus few social scientists would ask respondents whether they approve of murder because we would expect almost everybody not to. (Rossi and Rossi 1990, p. 160)

In population surveys, large variances in variables are sought to register the extent and shape of social controversies and to enable causal inferences. However, a survey of culture is intended to build a descriptive database regarding norms, and therefore lack of variability on every item is the ideal, since response variation confounds the delineation of norms. Surveys of cultures do seek variations, but across items rather than across respondents. For example, when describing a culture, the difference in the average evaluation of doctor versus the average evaluation of robber is the interesting variation, rather than differences in evaluations of doctor or robber by respondent A versus respondent B. Stable individual differences actually count as errors when surveying cultural norms because individual deviance obscures underlying uniformity. Consequently, usual notions of reliability no longer hold in culture surveys. In fact, as shown later in the book, an ideal item for assessing a norm would have zero reliability in the traditional sense!

An occasional complaint about surveys of culture is that the respondents providing the data are too few and are not selected randomly from general populations. However, this criticism—posed from the perspective of traditional survey methods—is tangential for ethnographic data gathering. The aim in ethnography is not to describe a population of individuals but, instead, to describe a culture that is being reproduced within some group. Properly chosen respondents are those whose responses are quintessential for their culture, and the more normative the respondents' beliefs and sentiments, the fewer of them are needed to obtain an accurate view of the culture. Whereas there is no notion of respondent goodness in surveying a population, other than representativeness of the sample as a whole, proficiency in the target culture is a key desideratum in choosing respondents for a survey of culture. Indeed, lack of cultural expertise is a legitimate basis for culling respondents or for assigning less weight to a respondent's answers, as we demonstrate later in the book.

The usual procedure in a survey of a population, of sampling individuals in a political unit, is inefficient in surveys of culture because desirable respondents for a survey of culture typically are not evenly distributed throughout a politically defined population. Rather, the best respondents for a culture survey are persons who reproduce the culture—the denizens of settings where the culture is being regenerated. For example, if interested in the middle-class culture that sustains the basic social institutions of American society (e.g., commerce, education, medicine, politics), a researcher would seek representative settings in which the activities of those institutions occur and question individuals at those sites about cultural matters. If interested in the culture that sustains black community life, the researcher would go to the homes, churches, and leisure venues where black culture is reproduced and question people at those sites. This emphasis on behavior settings in which culture is reproduced contrasts with the sampling frame in traditional survey research, where individuals themselves are the sampling units, even when geography is a practical consideration in sampling schemes for acquiring respondents.

So surveys of culture differ fundamentally from surveys of populations in at least three respects: (1) questions are asked about matters of agreement rather than about issues that generate diversity of response; (2) respondents are graded on the basis of their expertise; and (3) respondents are acquired by visiting settings where cultural reproduction takes place rather than by random sampling of people in a large geographic area delineated by political boundaries.

Since surveys of populations and surveys of cultures have distinctive goals and methods, one kind of study cannot be adapted easily to the purposes of another. A representative sample from a population rarely is simultaneously a sample that is culturally homogeneous and well inculcated in the culture of interest, so population surveys generally provide little information about the vast number of shared norms and understandings in the dominant culture. On the other hand, culture surveys aim at homogeneity in responses among those who are most knowledgeable about the culture, so they typically provide meager data about variations among diverse individuals within the society at large, and they are poor bases for explaining individual differences.

1.1 CASE STUDIES OF CULTURAL SURVEYS

Subjective culture—the knowledge and motivational structures a society provides for its indigenes—is the focal concern in this book. Surveys can be fielded to assess norms of material culture; for example, Chapin (1932) described how interviewers visiting homes could record household furnishings and equipment. Surveys can also assess some kinds of behavioral norms, such as patterns of pronunciation in American English (Labov, Ash, and Boberg 2006). However, the massive social science emphasis on cognition and emotion during the last quarter century has made the study of subjective culture a richly developed area, worthy of methodological consideration.

Important aspects of subjective culture include the categorization system that establishes culturally acknowledged realities, the implicit rules that interrelate cultural categories in ways that foster sensible decision making, and the sentiment system that orders cultural categories in terms of value, significance, and urgency. In the following case studies, I illustrate how social surveys have been applied in each of these areas.

1.1.1 American Regional English

In 1965, lexicographer Frederic Cassidy fielded a five-year survey of American variations in word usage that acquired such abundant data that the ultimate product, the *Dictionary of American Regional English* (*DARE*), still had not been completed at Cassidy's death in 2000. The first volume of *DARE*, 1,059 pages describing procedures and conventions and definitions for words beginning with the letters A through C, appeared 20 years after the survey began. Volumes 2 and 3, edited by Cassidy and Joan Hall and covering D through O, appeared in 1991 and 1996. Volume 4, edited by Hall and covering P through Sk, appeared in 2002. Volume V, covering the remainder of the alphabet, still was unpublished as of 2009. Another planned volume will contain a bibliography, maps, responses to the questionnaire, and such. (All of the volumes are available from Harvard University Press.)

The general concern of the *DARE* project was folk usage, or language learned at home from relatives or learned in the community from friends and associates (Cassidy 1985, p. xvi). There was no interest in words that could be considered Standard English as learned in schools or acquired from books, or as documented in conventional dictionaries; the project did not deal with technical or scientific words.

The questionnaire administered face to face to respondents in this project was massive. "If every question had elicited one response each time it was asked, there would have been 1,850,694 responses—but multiple answers brought the total closer to 2,500,000" (Cassidy 1985, p. xiv).

> Intended for use in personal interviews, the *DARE* Questionnaire (QR) begins with the neutral subject of time in order to allay possible suspicions of some hidden purpose on the part of the investigator. Next come weather and topography, equally neutral and safely concrete. Houses, furniture, and household utensils follow, with dishes, foods, vegetables, and fruits. And so the questions continue to more abstract topics: honesty and dishonesty, beliefs, emotions, relationships among people, manner of action or being, and so on—41 categories in all with a total of 1847 questions. (Cassidy 1985, p. xii)

Some questions described denotative targets and asked respondents to supply the words that they themselves used to refer to such an object. In this case, "the question, if properly phrased and understood, and the answer, if responsibly given, should ideally produce a reversed definition. For example: 'What do you call a container for coal to use in a stove?' Responses: *coal bucket, hod, pail, scuttle*, and so on. Reversed, this becomes a definition: *scuttle*, a container for coal to use in a stove. The method works relatively well for simple material objects like coal scuttles, less well for abstract things or emotional matters" (Cassidy 1985, p. xiii). Other questions named a category in Standard English and asked respondents to supply local synonyms. For example, one question asked what local names were given to the dragonfly (Cassidy 1985, p. lxix). Respondents gave 79 different replies to this question, among them *snake feeder, snake doctor, mosquito hawk, spindle, and ear-cutter* (Cassidy 1985, pp. xii–xiii). In a subset of questionnaires, respondents identified wildflowers from color photographs.

Completion of the entire questionnaire required 25 to 30 hours (Carlson 2001, p. 2). Since few respondents could commit such a huge block of their time, each questionnaire typically was divided among several respondents from the same community—2.8 respondents on average.

Besides obtaining respondents' answers to the questions, fieldworkers recorded individual information regarding the respondent: name, address, gender, race, age, education, amount of travel, chief occupations, associations, family background on both sides, and a description of the respondent's speech and attitudes toward language. Fieldworkers also prepared a brief description of the respondent's community. For most respondents, a tape recording was made of the respondent speaking freely for 20 minutes or more on a familiar topic and also reading a short story known to expose speech variations.

The fieldworkers mainly were graduate students who had some formal training in English and linguistics and who could make phonetic transcriptions; a few undergraduates and faculty members also worked in the field (Cassidy 1985, p. xiii). In total, the fieldworkers consisted of 51 men and 29 women. To deal with a conceivable biasing factor, both white and black fieldworkers interviewed black respondents; however, results appeared to be unaffected by the race of the interviewer (Cassidy 1985, p. xiv). Some of the fieldworkers have published interesting and valuable reminiscences of their fieldwork experiences in the *DARE Newsletter*, available at the *DARE* Web site (von Schneidemesser 2008).

The *DARE* project sampled communities rather than individuals, with relatively few communities being selected in sparsely populated Western states and many communities being selected in populous Eastern states; for example, New York City had 22 *DARE* communities (Carver 1985, p. xxiii).

> Communities were chosen in each state, the number proportional to the population and taking settlement history into account. . . . The aim was to choose relatively stable communities, distributed according to the states' composition, and communities of various types, so that the aggregate would reflect the makeup of each state's population. . . . *DARE* recognized as a "community" any group of people living fairly close to each other and sharing the same commercial facilities, social organizations, and the like. Even within metropolitan areas such communities, or subcommunities, exist with a sense of focus based on ethnic, religious, and other characteristics. Contrariwise, quite small independent rural communities, though close together, may keep themselves apart on similar grounds. (Cassidy 1985, p. xiii)

A total of 1,002 communities were selected in the 50 states of the United States, and one questionnaire was completed in each community, often by several different respondents. "Neither the choice of communities nor that of informants was randomized; on the contrary, the intention was to maximize the collection of materials by going to the places and people most likely to furnish the largest amount of appropriate data" (Cassidy 1985, p. xiv).

Fieldworkers chose informants, and while they attempted to balance gender, age, education, and race for their geographic area, inculcation into the community was a paramount consideration. "To qualify as a *DARE* informant, a man or woman had to have been born in the community represented or very close by, and could not have traveled or stayed away long enough for his or her speech to be affected. [Choices were made] with a deliberate weighting toward older people. Folk language is traditional, and older people remember many things that young ones have never heard of" (Cassidy 1985, p. xiv). Fieldworkers were advised to search for ideal informants by contacting "local churches, branch libraries, real estate offices, funeral parlors, gas stations, car dealers, and even neighborhood bars" (Carlson 2001, p. 2).

The 2,777 respondents in the study consisted of 1,368 men and 1,409 women. Sixty-six percent were 60 years of age and over (some over 90), 24 percent were middle-aged (40–59), and 10 percent were in the age range 18 to 39. Levels of education were: 3 percent less than fifth grade, 24 percent no more than one year of high school, 41 percent at least two years of high school, 31 percent at least two years of college, and 1 percent unknown. Racially, the respondents were 92.7 percent white, 6.7 percent black, with 0.3 percent each of American Indians and Orientals (Cassidy 1985, pp. xiii–xiv).

The *DARE* survey focused on identifying regional words or meanings, and in this sense it was not a survey of a single culture but of multiple subcultures in the United States, as manifested in America's variant Englishes. Language variations in the United States are not so distinctive as variant Englishes around the world (Bhatt 2001), but American usages do vary in ways that reflect the different social histories and interests of groups in different geographic settings.

Rather than specify the distinct regions of language usage by fiat, the *DARE* researchers took an empirical approach based on a novel form of mapping. "On [a] conventional map, the [*DARE*] communities are concentrated in the most populous areas of the country. ... At the same time the communities of the western states are widely scattered. The *DARE* map, by contrast, compresses the western states while expanding the more populous eastern states, creating a relatively uniform distribution of communities across the map. This makes it easier to see the clustering of communities where a given response is recorded" (Carver 1985, p. xxiii). In essence, *DARE* maps allow any viewer to partition response data visually into culture regions. "The *DARE* map is essentially a scatter diagram that economically illustrates degrees of clustering—that is, degrees of regionality" (Carver 1985, p. xxvii). The unique maps appear throughout the dictionary volumes, and a collection of them is planned for the final volume.

Cassidy (1985, p. xiii) remarked that the *DARE* project assembled unequaled data on living American speech, with the data containing a great deal of information on syntax that went unexploited in constructing the dictionaries. The data additionally could be analyzed to study American folk culture. Reading through the *DARE* questionnaire is a humbling experience regarding one's knowledge of American folk culture, which suggests that future analyses could examine the percentage of questions that respondents were able to answer satisfactorily, after eliminating answers with special codes,[1] in order to correlate the extent of folk knowledge with social variables such as age. Data from communities could be analyzed to identify fissures in American folk culture. While the *DARE* maps allow responses to a single question to be partitioned into homogeneous groups, a multivariate analysis of communities' responses to all questions, comparable to the Q-factor analyses discussed in Chapter 7, would partition communities and the content of folk culture simultaneously into general divisions, whether or not related to geography. Such analyses will eventually become possible when the plan to put the *DARE* corpus on the Internet is implemented.

1.1.2 Obligations to Kin

Sociologists Alice and Peter Rossi undertook a survey of family relations among people in the Boston metropolitan area in 1984 and 1985, publishing the results of their study in a 1990 book entitled *Of Human Bonding*. Their study focused on three main questions: How do individuals change over the life course? What determines variations in parent–child solidarity? How are obligations to help others normatively structured? Analyses relating to the last question—the one of interest here—generated findings about norms of help-giving in Boston, and probably elsewhere in America, which allowed the authors to portray how obligations vary from parents and children to siblings, grandparents, aunts, cousins, and on to nonkin such as friends, neighbors, and former spouses.

Data for the study of norms was obtained after an hour-long face-to-face interview that dealt with life-course matters and relations between parents and children. At the end of the interview the respondent took about 15 additional minutes to work through a questionnaire booklet containing 31 brief vignettes, preceded by a practice vignette that the interviewer used to instruct the respondent in the task. While the interviewer checked over answers that had been obtained in the interview, the respondent read through the vignettes and provided ratings about what level of help would be appropriate in each situation.

Rossi and Rossi (1990, p. 45) described items in the vignette questionnaire as follows:

[1] Interviewers entered special codes when they had to prompt for an answer and when they had doubts about the response; they also entered special codes when the respondent claimed no such word was used locally and when the respondent gave no response (Cassidy 1985, p. xiv).

A typical vignette might read:

> "Your married brother has undergone major surgery and will be disabled for a very long time. This problem is straining his financial resources. How much of an obligation would you feel to offer him some financial help?"
>
> A numeric rating scale follows this question, ranging from "0" (No obligation at all) to "10" (Very strong obligation). Two types of obligation ratings were built into the design: an *instrumental* obligation, in the form of financial help, as illustrated above, or an *expressive* obligation, for which the question was "How much of an obligation would you feel to offer him comfort and emotional support?"

The illustrative vignette above was in the category of crisis events involving a traumatic event in the life of that kinperson creating a situation of need that called for financial help or for comfort (Rossi and Rossi 1990, p. 162). Other vignettes dealt with celebratory occasions calling for recognition or appreciation, and ratings were made on scales relating to obligation to give something appropriate to the occasion or to visit (Rossi and Rossi 1990, pp. 162–164).

The 31 vignettes considered by a respondent were a small sample of the grand total of 1,628 vignettes presented in the survey. Different respondents got different sets of vignettes, each booklet being constructed uniquely as a random sample from the total population of vignettes. Ratings by different respondents later were pooled to analyze all vignettes together.

The 1,628 vignettes were created by conjoining 74 focal characters with various types of crisis or celebratory events. The focal character of each vignette was designated in terms of three dimensions: form of kinship, gender, and marital status.

> Using the respondent as the reference point, the kin designated ranged from four grandparents, each described as a grandmother or grandfather, whether in the maternal or paternal line and whether married or widowed, through the parental generation (e.g., mother father, uncle, or aunt), through the respondent's generation (siblings and cousins) to children and grandchildren. Each kin type was further specified by gender and marital status. Children and grandchildren were implicitly described as adults, the other kin being explicitly defined as such. To provide a contrast with the level of felt obligation toward nonkin, we also included "friends" and "good neighbors" and "former spouses" as designated categories. (Rossi and Rossi 1990, p. 164)

Eight types of crisis situations appeared in the vignettes, including major surgeries, serious personal troubles, losing everything in a household fire, and unemployment. Three types of celebratory events appeared: winning an award, having a birthday, and moving into a new place.

The overall survey involved face-to-face interviews with 1,393 respondents obtained via probability sampling procedures. The overall survey also obtained telephone interviews with 323 parents of main respondents and 278 adult children of main respondents, but those data were not used in vignette analyses. The vignette questionnaire was completed by 1,176 of the main respondents (84.4 percent).

About 49 percent of the respondents were age 19 to 40; about 9 percent were over 70 years of age. Fifty-eight percent were female. About 62 percent were currently married, and the rest were about evenly divided between never married and previously married. The respondents were overwhelmingly white (94 percent), Anglo (70 percent), and Christian (80 percent, two-thirds of whom were Catholic).

The various kinds of obligations—financial aid, emotional support, gift giving, visiting—conceivably could be complementary types of solidarity help-giving provided to different sets of relatives, but the Rossis found that the opposite is actually the normative rule (1990, pp. 169–170).

> Offering comfort and offering financial aid to a given kin tend overall to go hand in hand, respondents stating the relatives to whom they feel strongly obligated to give comfort and emotional support are also highly likely to be the same kin to whom they feel strongly obligated to offer financial help. . . . [Similarly] kin that induce a strong obligation to visit are also kin that evoke a strong sense of obligation to give gifts, and, correspondingly, kin to whom little or no obligation to visit is acknowledged tend also to be the kin to whom little or no obligation to send gifts is felt.

In essence, each type of kin has a general obligation strength that governs giving emotional support and offering financial aid in crises and that determines gift giving and visiting in happy situations. What differs among the various obligations is the threshold at which they dominate. For example, respondents typically felt more of an obligation to give comfort and emotional support than to provide financial aid, so comfort and emotional support end up being offered in a broader range of crises than is financial aid. For celebratory occasions, making a visit often is less obligatory than giving a gift, at least in the case of more distant relatives (Rossi and Rossi 1990, Table 4.7).

As one would expect, all types of normative obligations are strongest for one's parents or children, and strong obligations to these kin hold across all kinds of situational circumstances. After parents and children, siblings and grandchildren occasion the greatest normative commitment, then daughters-in-law, sons-in-law, and parents-in-law. Grandparents and stepchildren complete the innermost circle of kin. After that, normative obligations to friends often are as great as obligations to kin such as stepparents, nieces and nephews, aunts and uncle, and cousins. Normative obligations are least for ex-spouses, regardless of the type of obligation or the circumstances.

The Rossis classified focal characters in vignettes in terms of their *range*: primary kin (parents and children); first-order kin (connected through another relative: e.g., siblings, grandparents, and grandchildren); second-order-kin (connected through two other relatives: e.g., aunts and nieces); third-order-kin (connected through three other relatives: cousins); kin related by marriage (in-laws); step kin; friends and neighbors; and ex-kin. They also classified the vignette characters in terms of *depth*, or the other's generation relative to the respondent's: two generations up (grandparents); one generation up (e.g., parents, aunt); same generation (e.g., siblings, cousins, friends, ex-spouse); one generation down (e.g., children, niece); and two generations down (grandchildren).

Great variations in levels of obligations were found among the range categories, with little variation within each category. In the case of the depth categorization, they found higher obligations to descendent than to ascendant generations. Overall, the structure of kin relations, defined in terms of range and depth, largely determined obligations of all kinds, in all types of situations.[2] The Rossis concluded (1990, pp. 182–183):

> There is clearly a robust and consistent underlying structure to normative obligations to kin. Types of situational stimuli or types of obligatory responses show only minor differences, compared to the inherent structure determined by the degree to which respondents are related to the kinperson in the vignette. Obligations radiate out in lessening degrees from the high obligation primary kin, with greater obligations toward descendents in all categories of kin than to ascendants. Secondary affinal kin, acquired through marriage or remarriage, generally evoke greater obligations than distant consanguineal kin. Friendship involves considerable obligation to provide social-emotional comfort, on a par with secondary blood kin, but . . . friends do not stimulate as high an obligation to provide financial aid as do blood and affinal kin.

The Rossis used the range and depth categorizations to compute regression equations predicting the average obligation felt toward each type of vignette character. They summarized these analyses as follows (Rossi and Rossi 1990, pp. 183–185):

> The most striking findings are that the handful of coded structural features of kinship accounted for almost all of the variation in the strengths of obligations to the kin in question. The R^2 values for the four kinds of obligation . . . range from .89 to .94, which are considerably higher than ordinarily found in social science research. These findings further indicate that there is a very robust normative structure to American kinship: Obligations to kin vary in a lawful and regular way according to the position of the kinperson in question vis-à-vis ego. . . . Primary kinship relationships are the most obligating, with obligation declining rapidly as the number of links between ego and kinperson increase. Affinal kin are more obligating than step kin, but both are less so than consanguineal kin.

[2]Malone (2004) found that the kinship structure similarly determines sentiments about relatives when sentiments are measured in the manner described in Chapter 2 of this book.

> Finally, kinship relationships that go down the generational ladder are more obligating than those involving the same or ascendant generations.... Note also that female kin are slightly more obligating than male kin, a generalization that does not hold for visiting, but does hold for the other obligation types.

In further analyses, the Rossis found that obligations also were determined somewhat by the nature of the person in the connecting link between ego and the focal character. Kin who were related through siblings had lower priority than those related through a parent, child, or spouse; and the obligation level increased when the connecting link between ego and alter was a woman (Rossi and Rossi 1990, p. 207).

Considerable consensus among respondents was necessary to reveal such structured results, but agreement was not perfect concerning levels of obligation to various kin in a range of circumstances. The least variance in ratings occurred for children and parents, the most variance for grandparents, stepparents, and stepchildren, with other kinds of kin and nonkin ranged in between. The Rossis conducted numerous analyses to find causes of variations in obligation ratings.

The ethnicities of respondents had some small impacts on their obligation ratings (Rossi and Rossi 1990, pp. 239–241). Irish and Jews gave somewhat higher obligation ratings than did respondents of British extraction. Relative to respondents of other ethnicities, blacks, Asians, and Portuguese gave relatively high ratings to distant kin.

Respondents' age was an important factor (Rossi and Rossi 1990, pp. 220–225). In general, younger respondents rated all kinds of obligations as stronger than did older respondents. Age 50 was the main breakpoint; after age 50 obligation ratings declined steadily with every decade of aging. Moreover, older respondents rated obligations as closer to zero for every type of kin—primary, secondary, and distant—and even for nonkin. The Rossis could provide no explanation of this phenomenon, although they noted that it was consistent with other data in their survey regarding help-giving between parents and their adult children.

Education explained variation in ratings to some degree, with college-educated respondents rating all kinds of obligations as stronger than did high school graduates, who in turn made ratings farther from zero than did those without high school degrees (Rossi and Rossi 1990, pp. 226–230). The education effect acted independently of the age effect; education accounted for less variation in ratings than did age.

Emotional troubles in a respondent's family of origin (e.g., alcoholism, mental illness, child or sexual abuse) lowered the respondent's obligation ratings for all kinds of kin. On the other hand, many other kinds of early difficulties, such as family quarrels or problems with schools, actually increased respondents' obligation ratings. "The joint effects suggest that adversities that have their roots in events beyond the control of parents—physical illnesses, unemployment, the behavior of children—lead to stronger kin bonding. In contrast, troubles that have their roots in the behavior of parents (or other adults) lead persons who experience such adversities in their childhood to become adults with lower levels of kin obligations generally" (Rossi and Rossi 1990, p. 235). A respondent's own divorce also was associated with lower obligation ratings for all kinds of kin (Rossi and Rossi 1990, Table 5.14).

The Rossis summarized their findings as follows (Rossi and Rossi 1990, p. 246):

> Virtually all of our respondents feel some degree of obligation to kin as prescribed in our kinship norms. However, some feel strong obligations to kin and in others there are weaker ties to the kindred. Although we cannot explain all of the differences from respondent to respondent, it is also apparent that their childhood families can set down strong or weak patterns, the former by cohesive families presided over by affectionate parents. As adults, the strength of obligations is influenced positively by being better educated, having a strong sense of duty in a variety of roles, and by being an outgoing expressive person.

As will be seen in Chapter 6, many of the Rossis' findings parallel findings in other studies relating to respondents' cultural inculcation.

1.1.3 African-American Sentiments

As a part of a larger project devoted to studying the cross-cultural validity of three dimensions for measuring sentiments and to obtaining comparative data on sentiments in more than two dozen cultures (see Chapters 2 and 3), a group of researchers led by Daniel Landis surveyed a broad range of sentiments held by African-American youths living in segregated areas of Chicago during the 1970s. They reported their procedures and a selection of findings in a book chapter (Landis et al. 1976). Additionally, a listing of black sentiment measurements for 611 concepts was sold as a computer printout at the University of Illinois Bookstore during the 1970s (Landis and Saral 1978); see Chapter 3 for more details. The computer printout introduced the African-American study as follows: "The data in this atlas were gathered from lower-class Black male youth in Chicago during 1973–74. Data gathering sites were from high-schools located in center-city and the southern schools. The concepts used as well as the scales were based on the patois in use during the same period (actually 1971–74)."

A goal of the study was to determine if black sentiments reflected a culture different than the dominant white culture recorded in sentiment measures obtained from whites during the 1960s. Landis et al. (1976, pp. 50, 55) noted that some researchers too readily assumed that black and white cultures were the same:

> Most American researchers seemed to forget that the United States itself is a very heterogeneous culture. There tended to be an implicit assumption that the

> original American English data could be applied willy-nilly to various American subgroups, including black Americans. . . . The statement of the problem here does not mean that we accept (or reject) the hypothesis of black cultural uniqueness. But given the anthropological, historical, and sociological data, such a hypothesis is at least tenable.

Determining if black sentiments reflected a separate culture required Landis and his colleagues to reproduce methodological procedures used in foreign locales, aimed at preventing ethnocentric biases. Only by treating the black sentiments as culturally disparate could it be determined whether or not they were. Accordingly, Landis and his colleagues treated African-American Vernacular English (AAVE) as a language in its own right.

The process of obtaining measurements of black sentiments free of white biases began with a list of 100 nouns common to all of the languages in the larger cross-cultural study (Osgood, May, and Miron 1975, Table 3-1); the list included words such as *house, girl*, and *meat*. These words were translated to AAVE by 10 translators,[3] "whose ages ranged from late adolescence to early twenties, [and who] had grown up in a lower SES urban ghetto environment. All were training to become teachers in such a setting" (Landis et al. 1976, pp. 57–58).

As Landis et al. (1976, p. 45) remarked: "The words used [in AAVE] are sufficiently at variance from their use in standard American English to provide the potential for serious misunderstandings between some blacks and whites." For example, translating to AAVE turned a house into a crib, and a girl into a little-moma (Landis and Saral 1978, Table 9-C). Of course, some words (e.g., *meat*) did stay the same.

The first function of the translated nouns was to serve as stimuli eliciting a range of frequently used adjectives in AAVE. The nouns were presented to 100 black high school students, and these respondents were told to give common adjectives used on the street with each noun, in frames like A crib is _, or The _ crib. "The testers were black males, and the instructions that they gave to the *S*s stressed that the adjectives should be those likely to be used by the 'black and beautiful people' " (Landis et al. 1976, p. 58). This resulted in the naming of hundreds of words, the top 200 of which ranged from frequent responses such as *bad, good, big, nice, cool, eat, black*, and *together* to relatively low-frequency responses such as *new, scared, do-your-thing, happy, good-looking, fun*, and *chicken* (Landis and Saral 1978, Table 2). Sixty adjectives for further study were chosen for their high frequency of usage, their usage with a variety of different nouns, and their complementarity with other adjectives.

[3]The first three steps of the procedure—translating nouns, eliciting adjectives, and selecting adjective opposites—were performed by New Jerseyites, because that was the original intended locale for the study, but the data were deemed equivalent to data that might be gathered in the final study locale of Chicago.

The purpose of eliciting adjectives was to obtain modifiers that could be turned into anchors for bipolar scales. The process of scale construction was continued by specifying opposites of the adjectives obtained. "The original group of translators was called together and asked to provide the opposites for the 60 adjectives. . . . Again, the emphasis was on the opposites that would be used in everyday street conversation" (Landis et al. 1976, pp. 58–59). Matching adjectives with their opposites and including a set of checkmark positions in between generated a set of 60 bipolar adjective scales (Landis et al. 1976, Table 1). The adjective anchors of these scales ranged from the familiar in standard American English (e.g., *fast–slow*) to pairs that might seem strange (e.g., *straight–stone*).

The 60 bipolar adjective scales were used to rate each of the nouns that had been used previously to elicit adjectives. Rating all nouns on all scales would have been too big a task for any one respondent, so booklets were made up with a random selection of 10 nouns for each booklet. "In constructing the booklets, the 60 scales were randomized on a page in terms of order as well as position (left versus right) of the 'positive' end of the scale. Each booklet then consisted of 20 pages (two pages for each concept, with each page containing 30 scales)" (Landis et al. 1976, p. 59). The number of booklets was doubled by duplicating them with scale order reversed.

Ratings of nouns on the scales were obtained from black youths in Chicago (Landis et al. 1976, p. 59).

> Each of the test booklets was then administered to at least 20 black adolescent pupils in the ghetto area of the West Side of Chicago. Although it was originally planned that each *S* would complete an entire booklet (that is, make a total of 600 judgments), this proved impractical given school time constraints and a high absentee rate at test sites. Therefore each *S* completed the rating of at least one concept on all 60 scales. Some, of course, did more.

Factor analyses determined the dimensionality of the scales. The data matrix had 60 columns and a row for each of the rated nouns. Each cell entry was the mean of all ratings that had been made of the noun representing the cell's row, on the scale representing the cell's column.

Three factors, or dimensions, accounted for 52 percent of the variance in the cells (Landis et al. 1976, p. 61). The first dimension—identified as Evaluation—related to ratings on the *good–foul*, *all right–mad*, *hard up–straight*, and *peaceful–ferocious* scales. The second dimension—identified as Potency—related to ratings on the *large–small*, *big–small*, *big–little*, and *wide–frail* scales. The third dimension—identified as Activity—related to ratings on the *active–passive*, *free–tied down*, *fast–slow*, and *loose—tight* scales.

Landis and colleagues conducted a bicultural factor analysis that analyzed mean ratings of the 100 nouns on black scales used by blacks along with mean ratings on white scales used by whites (Landis and Saral 1978, Table 5). The bicultural analysis revealed whether factors emerging in the black data were similar to factors emerging in white data. Two correlated Evaluation factors emerged for blacks. One of these, with the scales *clean–nasty*, *jam–bad scene*, *good–foul*, and *together–wrong*, was the same as the Evaluation dimension defined by white ratings. The other black evaluation scale was defined by scales such as *hip* versus *dumb*, *cool* versus *silly*, and *hip* versus *lousy*. The Potency factors for blacks and whites aligned, and so did the Activity factors. Landis and his colleagues selected scales to measure each of the three dimensions common to blacks and whites and computed factor scores for each of the nouns that had been rated. The results were presented in a table (Landis et al. 1976, Table 3).

Landis and colleagues compared their measurements of black and white sentiments with regard to selected concepts in the general areas of material possessions, confrontation, personal relations, quality of life, and ecosystem distrust. Their interpretative approach was one of cultural equivalence, legitimating middle-class family norms, and stressing similarities between blacks and whites.

> Our feeling from the data presented in the previous pages is that not only do blacks value the same goals, relationships, and ideals that whites do, but in many cases they value them more. Where the differences occur, they seem to be related to perceptions of the amount of effort necessary to achieve those goals and the potency of those aims in changing one's life. In other words, blacks want the same things whites do, but they don't believe that (a) they can achieve them, (b) if they could, their lives would be significantly improved, and (c) they would be engaged in anything less than a constant struggle to maintain those things they do achieve. (Landis et al. 1976, p. 78)

Later Landis and Saral (1978, Table 6) factored mean ratings of the culturally common nouns on scales from 23 cultures simultaneously—a pancultural factor analysis. The intent was to see if the dimensions of black ratings corresponded to rating dimensions that had been found to be present in all of the cultures studied in the overall project (Osgood, May, and Miron 1975, Chapter 4). Black dimensions did indeed emerge that aligned with cross-culturally universal dimensions. Sets of black scales were selected to measure each of the pancultural dimensions, and these scales were then used to rate several hundred concepts that also were rated in all locales within the larger cross-cultural project. This part of the cross-cultural project was described in general terms by Osgood (1974, p. 1): An *Atlas of Affective Meanings* was to be constructed from mean ratings of 620 common concepts rated by respondents in all of the cultures in the study in order to provide a basis for exploring "subjective culture—values, feelings, and meanings—as it is expressed in language." The 620 concepts sampled many areas, from universal notions prevailing in large societies, such as colors and numbers, to concepts arising in everyday life related to kinship, foods, animals, technologies, and other matters considered routinely by ethnographers.

The Landis and Saral (1978) report presenting black factor scores for 611 concepts was fallow for three decades. However, Sewell and Heise (2009) returned to the data and conducted cross-cultural analyses comparing the black sentiments, white American sentiments, and German sentiments. Sewell and Heise found that while black and white mean factor scores correlated positively on the three dimensions of Evaluation, Potency, and Activity—0.65, 0.38, and 0.03 respectively—these correlations were less than the correlations between white Americans and Germans—0.83, 0.66, and 0.58. That remained the case even when correlations were computed just with concepts that did not translate into different words in AAVE and when those correlations were corrected for unreliability. Summarizing, Sewell and Heise said: "We conclude that during the 1970s, sentiments in some Black groups were distinctive enough to be treated as a parallel subjective culture co-existing with the White subjective culture—as different from White culture as White culture was different from the culture of another nation."

1.1.4 Observations on the Cases

The *DARE* study attempted to distinguish, delineate, and compare different subcultures of American folk English, whereas the other two studies focused on homogeneous populations in which respondents could be assumed more or less inculcated with uniform norms. One consequence was that the *DARE* study had to employ a method of factoring different groups apart, the ingenious use of population-weighted maps. Another consequence was the substantial amount of time required for analysis. The *DARE* study spanned 20 years from data collection to the first volume of lexicographic results, with the final volume expected 44 years after data collection. Part of the delay in the *DARE* study was due to its qualitative nature: Every definition had to be constructed uniquely. However, another consideration was the study's cross-cultural aspect: Cross-cultural studies involve problems of synchronization and matching of responses for comparisons, beyond regular data analysis (see Harkness, van de Vijver, and Mohler 2002). The time required for analysis contrasts with the Boston kinship study, which was completed in a half-dozen years from data collection to book publication. The study of black sentiments culminated in a book chapter within a few years of data collection. On the other hand, the larger cross-cultural project of which the black sentiments study was a part did take years to produce results. A first volume of results was published about 10 years after data collection began, but a planned second volume devoted to cross-cultural comparisons was still in process when the project director (Charles Osgood) fell ill, and that volume never appeared.

Some uniformity is evident in all three examples of culture surveys. In all three cases, data were collected by visiting settings where the culture of interest was being reproduced day by day. The *DARE* investigators were explicit about this, designing their project so that it would sample communities where different varieties of American English might be spoken, and giving instructions to interviewers to visit locales where folk conversations often take place. The study of African-American sentiments used urban black neighborhood high schools for its data-collection venues. Interactions among teenagers are settings where a unique black culture is likely to be found, as reflected in Labov's (1972, p. 257) remark that "the most consistent vernacular is spoken by those between the ages of 9 and 18." Although the study of family relations in Boston used a random probability sample of adults from the Boston Metropolitan Statistical Area, this was implemented by drawing an area probability sample of housing units in the Boston area. Thus the core sampling procedure focused on households where family culture was reproduced.

Often, seeking out settings where the culture of interest is being reproduced is associated with data collection in a single area, as in the Boston study of families, and the assessment of sentiments among Chicago blacks. However, when multiple areas are sought, culture researchers use population-weighted geographic sampling, as in the *DARE* study. [Similarly, Berk and Rossi (1977, p. 9) used numbers of prisoners as a weighting factor in selecting states for studying normative aspects of how elites were involved in prison reform).]

The *DARE* study and the study of black sentiments manifestly selected respondents for their levels of inculcation into cultures of interest. In the *DARE* study, respondents had to be natives of their area, preferably elders, with no diminishment of their cultural inculcation through extensive outside experiences. In the black sentiments study, urban black youths were employed exclusively as respondents and as language experts making study-related decisions. The focus on cultural expertise is less evident in the case of the study of family norms, but even there the respondents, all of whom were selected from Boston households, would have been participating in the reproduction of the family institution as they made daily decisions regarding their own family responsibilities. The sampling design omitted homeless and institutionalized persons, who would have been least likely to provide expert judgments regarding family obligations.

In all three of the case studies presented above, very long questionnaires were partitioned into relatively small subquestionnaires for administration to respondents. The long questionnaires reflected a basic consideration in cultural studies: A considerable range of material has to be covered to provide systematic treatment of a culture, as opposed to entertaining samples of cultural curiosities in the manner of newspaper feature articles. The need to subdivide the long questionnaires arises from the limited time and patience of respondents. The *DARE* study plumbed respondents' upper limits of these resources with interviews averaging 10 hours; that such long interviews were possible speaks to respondents' interest in their folk culture as well as to the adeptness of the *DARE* interviewers. The black sentiments study skirted respondents' minimum involvement, with some respondents providing ratings of only a single concept. The tediousness of repeated ratings on the same scales, especially for adolescents, probably was the key factor in such limited participation, although limited literacy of some respondents might also have been a factor slowing their use of a paper-and-pencil questionnaire.

1.2 PREVIEW

The rest of this book deals with surveys of culture, but not equally with all types of culture surveys nor with all issues arising in such surveys. The focus is narrowed to some methodological issues in surveying cultural sentiments such as those assessed in the case study regarding American blacks and in the use of vignettes to survey unconscious but normative responses to situations such as those considered in the case study regarding kinship. A primary reason for narrowing the focus this way is that sentiment and vignette studies are the aspects of subjective culture that I myself have researched for decades and that I know best. Beyond that, sentiments (including attitudes) and normative processing of situations prior to decision making are among the most actively researched topics in the social sciences, so the book's focus is a justifiable hub of many research interests.

Surveys of sentiment norms can arise in studies of various kinds of social partitions: for example, race, gender, geography, nationality, ethnicity, education, academic discipline, occupation, and leisure pursuits. In the past, data-collection projects have assessed sentiment norms for several thousand concepts in multiple nations and languages (see Chapter 3), and additional surveys of this kind are in progress. One driving force behind these surveys is affect control theory (Heise 2007; MacKinnon 1994; Smith-Lovin and Heise 1988), which argues in essence that social interaction, institutional roles, emotions, and other social phenomena are a function of people maintaining cultural sentiments. The sentiment surveys, combined with affect control theory's mathematical model and computer simulation software, allow predictions to be generated about specific social processes in various cultures. Applications have been made to courtroom processes (Robinson, Smith-Lovin, and Tsoudis 1994; Tsoudis 2000a, 2000b; Tsoudis and Smith-Lovin 1998, 2001), business organizations (Schneider 2002; Schröder and Scholl 2009; Smith 1995), and international relations (Heise 2006; Heise and Lerner 2006). The surveys also produce culture databases that can be considered with databases for other cultures in comparative analyses (Heise 2001a; Ragin 1987).

The *DARE* study represents a class of studies devoted to recording culturally shared knowledge. I do not focus on this kind of work because it has been examined so thoroughly in psychological anthropology by the ground-breaking originators of the culture-as-consensus approach and their collaborators (Batchelder, Kumbasar, and Boyd 1997; Batchelder and Romney 1988; Romney 1994, 1999; Romney, Batchelder, and Weller 1987; Romney, Weller and Batchelder 1986; Romney and Weller 1984; Romney et al. 2000; Weller 1987; Weller, Romney, and Orr 1987). One of their publications

(Romney, Weller, and Batchelder 1986) is the most highly cited article ever to appear in *American Anthropologist* (Batchelder 2009). Their model has been applied in studies of folk medical beliefs, parental sanctions for rule breaking, judgment of personality traits, semiotic characterizations of alphabetic systems, occupational prestige, causes of death, strategies to control graffiti, national consciousness in Japan, and social network data.

Chapter 2 of the book reviews the development of sentiment measurement procedures. Key theoretical ideas regarding sentiments were in place at the beginning of the twentieth century. However, the development of empirical measurement procedures took another 50 years, with a mammoth cross-cultural study being a key component in the development. The last half of the twentieth century and the turn of the twenty-first century saw the incorporation of computers and the Internet into sentiment-measurement methodology.

Chapter 3 describes repositories of sentiments that can be used in cross-cultural studies. This line of research began in the 1950s and burgeoned from about 1970 onward. Repositories have been assembled in a score of nations, and some of these repositories provide quantitative measurements regarding thousands of sentiment norms. Chapter 3 also describes in detail the sentiment study that provided data used in analyses for Chapters 6 through 8.

Chapter 4 describes a vignette method for assessing norms of unconscious cultural processing. The first part of the chapter focuses on how vignettes have been used to explore the generation of preferences, as in the Boston study of kinship described earlier. Other studies of this kind are also considered in the review. The second part of the chapter describes how vignettes have been used in multiple societies to identify norms in forming impressions from events and from other kinds of observations. The chapter includes an appendix offering guidance for designing impression-formation studies.

Chapter 5 expands Romney, Weller, and Batchelder's (1986) culture-as-consensus model in psychological anthropology to culture surveys in general, dealing with sentiment measurements on quantitative scales instead of just data with dichotomous categories, which has been typical in culture-as-consensus research. A review of contemporary research on errors in surveys is followed by the formulation of a mathematical model that undergirds the subsequent three chapters.

Chapter 6 focuses on respondents' adequacy as informants about norms, and more generally on their levels of cultural inculcation. Performance measures obtained during the rating task are found to identify respondents whose contributions actually undermine the assessment of norms. An enculturation index is used to predict inculcation levels from background information, and it is found that respondents' social characteristics are associated with their levels of enculturation, although the correlations are only modest.

Chapter 7 examines the empirical tenability of assumptions involved in culture-as-consensus methodology. Factor analysis applied to respondents' evaluations of concepts reveal that ratings by different persons form a single factor, supporting the assumption that respondents are normatively homogeneous. A dominant factor also characterizes ratings of a concept's potency levels and activity levels, but in both of these cases a second appreciable factor also appears, because some respondents transform a concept's evaluation into an assessment of the concept's potency and activity. Additional analyses demonstrate that factor analysis of sentiment data cannot be used to uncover subcultures, because adherents of a subculture have special sentiments for so few concepts relative to the total number of concepts in a culture.

Chapter 8 analyzes test–retest data for a few selected stimuli, plus one-time sentiment measurements for a large set of stimuli, in order to partition rating variances into cultural, individual, and error components. Converting these variances into measures of reliability reveals two key findings. First, ratings of a single concept in a normative study have fairly low reliability when reliability is conceived in a traditional manner, precisely because respondents' sentiments are normatively shaped and therefore similar to one another. Second, measurement reliability is substantially higher when reliability is conceived as the proportion of rating variance that is explained by the meanings of different concepts. Additional analyses show that the reliabilities of culture assessments increase dramatically when aggregating data from multiple respondents.

Chapter 9 contrasts culture surveys with traditional ethnographic studies and with traditional survey research studies. Despite some parallels, it is shown that culture surveys are a distinctive methodological approach that cannot be reduced to either ethnography or traditional survey research. An examination of sentiment data used in this book reveals that ratings of large samples of concepts are largely governed by meanings of the concepts rather than by idiosyncratic views of respondents, thereby emphasizing the extent to which traditional survey research studies focus on concepts that are in cultural play. The chapter also develops some guidelines for optimizing data quality in surveys of cultural sentiments.

1.3 CHAPTER HIGHLIGHTS

- In culture surveys, questions are asked about matters of agreement, so relatively few respondents need be surveyed to establish any particular norm. The more normative the respondents' beliefs and sentiments, the fewer of them are needed to obtain an accurate view of the culture.
- In culture surveys, a considerable range of norms has to be assessed systematically, so large samples of respondents may be required, with each respondent reporting on just a portion of the norms.
- In surveys of culture, lack of variability on every item is the ideal, since response variation confounds the delineation of norms. Surveys of cultures do seek variations in response, but across items rather than across respondents.

- The best respondents for a culture survey are denizens of settings where the culture is being regenerated. Thus, ideally, respondents are acquired by visiting settings where cultural reproduction takes place.
- Surveyable aspects of subjective culture include the categorization system that establishes culturally acknowledged realities, the implicit rules that interrelate cultural categories in ways that foster sensible decision making, and the sentiment system that orders cultural categories in terms of value, significance, and urgency.

2 Measuring Sentiments

The nineteenth-century father of experimental psychology, Wilhelm Wundt, saw affective states, or feelings, as distributed along bipolar spans characterized by adjective opposites. The number of affective states, and the contrasts for characterizing them, are practically infinite, Wundt said, yet three basic directions organize much of the multifariousness.

> In this manifold of feelings, made up, as it is, of a great variety of most delicately shaded qualities, it is nevertheless possible to distinguish certain different *chief directions*, including certain affective opposites of predominant character. Such directions may always be designated by the *two* names that indicate their opposite extremes. Each name is, however, to be looked upon as a collective name including an endless number of feelings differing from one another.
>
> *Three* such chief directions may be distinguished; we will call them the direction of *pleasurable* and *unpleasurable* feelings, that of *arousing* and *subduing* (exciting and depressing) feelings, and finally that of feelings of *strain* and *relaxation*. Any concrete feeling may belong to all of these directions or only two or even only one of them. The last mentioned possibility is all that makes it possible to distinguish the different directions. The combination of different affective directions which ordinarily takes place, and the ... overlapping of feelings arising from various causes, all go to explain why we are perhaps never in a state entirely free from feeling, although the general nature of the feelings demands an indifference-zone. (Wundt 1897, pp. 82–83)

Each direction defines two distinctive qualities and many levels of intensity. "The middle point between these two opposites corresponds to an absence of all intensity" (Wundt 1897, pp. 82–83), while very extreme intensities become emotions (Wundt 1897, p. 169). Thus, according to Wundt, three dimensions of affect range outward from a neutral point: an evaluative dimension characterized in terms of pleasurable versus unpleasurable, an activation dimension of arousing versus subduing, and a muscularity dimension of strain versus relaxation. A perception may make one feel pleasurable, excited, and taut; another perception may make one feel unpleasant, depressed, and soft; still another perception may make one feel unpleasant, aroused, and mobilized—and so on, through eight different combinations. Each feeling has a level of

Surveying Cultures: Discovering Shared Conceptions and Sentiments, By David R. Heise

intensity on each dimension, and some feelings are at the zero point of one or more dimensions. Thereby the number of distinguishable affective states is huge.

Somewhat later, in psychology's first textbook of social psychology, William McDougall (1908, p. 437) promoted the notion of a *sentiment* as "a system in which a cognitive disposition is linked with one or more emotional or affective conative dispositions to form a structural unit that functions . . . as one configuration or Gestalt." McDougall's concept of sentiment expanded the domain of affective association beyond Wundt's affectively laden sensations to cognitions of all kinds, including ideas and concepts acquired from culture: "A sentiment is an enduring structure within the total structure of the mind," McDougall said (1908, p. 436). Although renowned for his theory of instincts, McDougall saw sentiments as the basis of sophisticated human action (1908, pp. 438–439): "In the man of developed character, very few actions proceed directly from his instinctive foundations: perhaps an occasional start of fear or sudden gesture of anger; but all others proceed from his sentiments, that is to say, from the complex interplay of the impulses and desires springing (as regards their energy) from the conative dispositions incorporated in his sentiments, and guided (as regards the lines of their expression and action in striving towards their goals) by the whole system of acquired knowledge both of the object of the sentiment and of its relation to the world in general."

Putting the ideas of Wundt and McDougall together at the beginning of the twentieth century would have produced a perspective on affect and cognition similar to that which is sustaining considerable social research at the beginning of the twenty-first century. Namely, cultural entities are internalized in people's minds not only with cognitive meaning schemes, but also with affective associations that vary along three bipolar dimensions: goodness versus badness, weakness versus powerfulness, and quiescence versus activation; and the affectivity of cognitive concepts is the foundation of individual motivations in interpersonal and institutional activities (MacKinnon 1994; Scholl 2008).

A synthesis of Wundt and McDougall did not happen in the early twentieth century. Instead, psychological and social researchers turned their attention to attitudes. Like the construct of sentiment, the construct of attitude refers to a cognitive, affective, and behavioral (conative) complex, but the attitude construct collapses the affective realm to the single dimension of evaluation. This simplification was auspicious scientifically. Whereas Wundt gained his insights by introspections and McDougall gained his by scholarship and intuition, the new approach to affectively laden cognitions initiated the era of empirically grounded developments.

The astonishment and excitement regarding a scientific approach to mental contents can be sensed in the title of a landmark publication of the era, "Attitudes Can Be Measured" (Thurstone 1928). In this article, Thurstone (1928, pp.530–531) confronted potential incredulity directly.

> An attitude is a complex affair which cannot be wholly described by any single numerical index. For the problem of measurement this statement is analogous to the observation that an ordinary table is a complex affair which cannot be wholly described by any single numerical index. So is a man such a complexity which cannot be wholly represented by a single index. Nevertheless we do not hesitate to say that we measure the table. The context usually implies what it is about the table that we propose to measure. We say without hesitation that we measure a man when we take some anthropometric measurements of him. The context may well imply without explicit declaration what aspect of the man we are measuring, his cephalic index, his height or weight or what not. Just in the same sense we shall say here that we are measuring attitudes. We shall state or imply by the context the aspect of people's attitudes that we are measuring. The point is that it is just as legitimate to say that we are measuring attitudes as it is to say that we are measuring tables or men.

Thurstone argued that "since in ordinary conversation we readily and understandably describe individuals as more and less pacifistic or more and less militaristic in attitude, we may frankly represent this linearity in the form of a unidimensional scale" (Thurstone 1928, p. 538). Thurstone's method was to array verbal expressions of attitude (opinions) along a continuum with the help of a panel of judges. Such a scale then made it possible to "measure the subject's attitude as expressed by the acceptance or rejection of opinions" (Thurstone 1928, p. 533). For example, a scale for measuring attitude toward the church (Chave and Thurstone 1929, pp. 60–63) had at one end the items "I think that the church is a parasite on society" and "I think the teaching of the church is altogether too superficial to have much social significance"; toward the middle the item "I believe in what the church teaches but with mental reservations"; and toward the other end the items "I enjoy my church because there is a spirit of friendliness there" and "I believe the church is the greatest influence for good government and right living." Many people who endorse the opinions at either end would not endorse the opinions at the opposite end or at the middle, and many people endorsing the middle opinion would not endorse opinions at either end.

The Thurstone method of selecting opinions that characterize specific positions along an attitude continuum later was supplemented by other kinds of scales (see Heise 1974). Cumulative scales use items that are endorsed by everyone on one side of a particular level of attitude: for example, "I believe Mexican immigrants should be allowed to work in the United States," "I would welcome a Mexican immigrant in my neighborhood," and "I'm agreeable to one of my children marrying a Mexican immigrant." Linear scales compute the means of respondents' degree-of-agreement ratings regarding extreme opinions: for example, "Abortion of a human embryo is murder"; "Women have the right to abortion as a method of birth control." Another method, having respondents rate attitude topics on bipolar scales anchored at each end by adjective extremes, emerged at mid-twentieth century, and development of this method returned interest to all three dimensions of affect.

2.1 DIMENSIONS OF AFFECT

Synesthesia is a psychological phenomenon in which sensations in one sense domain cause sensations in a different sense domain. Psychologist Charles Osgood in his textbook on experimental psychology described synesthesia research in which subjects listened to excerpts of classical music and reported what moods the music aroused in them or what colors they would associate with the music. Normative responses were found for both moods and colors. Other subjects were asked to translate mood adjectives into colors, without any music, and "even *more* consistent relations were obtained, suggesting that the unique characteristics of the music had, if anything, confused the purely verbal or metaphorical relations between color and mood" (1953, p. 645). Impressed by how colors and moods—specified in terms of appropriate adjectives—translated readily from one to the other, Osgood conducted his own research with synesthesia and with adjectives specifying identity stereotypes, and these studies led him to a program of research on the dimensionality of meaning that occupied much of his later career.

Osgood perceived his work on meaning as resting on three basic assumptions (Osgood 1953, p. 713, italics removed).

> (1) The process of description or judgment can be conceived as the allocation of a concept to an experiential continuum defined by a pair of polar terms. . . . (2) Many different experiential continua, or ways in which meanings vary, are essentially equivalent and hence may be represented by a single dimension. It was this fact that was borne in on us in the synesthesia and stereotype studies. In the latter, for example, the descriptive scales fair–unfair, high–low, kind–cruel, valuable–worthless, Christian–anti-Christian and honest–dishonest were all found to be intercorrelated .90 or better, as used in judging social concepts. It is this fact about language and thinking that makes the development of a quantitative measuring instrument feasible. If the plethora of descriptive terms we utilize were in truth unique and independent of one another—as most philosophers of meaning seem to have supposed—then quantitative measurement would be impossible. (3) A limited number of such continua can be used to define a semantic space within which the meaning of any concept may be specified.

Osgood used bipolar adjectives to allocate concepts into semantic space and thereby to assess meanings.

> Presented with a pair of descriptive polar terms (e.g. *rough–smooth*) and a concept (e.g. LADY), the subject merely indicates the direction of his association (e.g. LADY–*smooth*) and its intensity by either the extremeness of his checkmark on a graphic scale or the speed of his reaction in a reaction-time device. The distribution of his judgments on a standardized series of such scales serves to differentiate the meaning of this concept from others; for this reason this measuring instrument has been called a "semantic differential." (Osgood 1953, p. 713)

Ratings of concepts on a series of bipolar scales produced quantitative profiles that allowed the concepts to be differentiated from one another. For example, Osgood (1953, Figure 218) found that subjects rated the concept *eager* and the concept *burning* both as very active, but *burning* was rated as hot while *eager* was rated as neither hot nor cold, and *burning* was rated as a bit bad while *eager* was rated as quite good.

Osgood's assumptions about meanings presumed that positioning a concept on a few basic dimensions would largely account for ratings on a multitude of specific bipolar scales. Osgood proposed to prove this by obtaining ratings of multiple concepts on multiple scales, and showing through the use factor analysis that much of the variance in the ratings could be accounted for by a few latent axes. That was a substantial challenge in the 1950s: Factor analysis was cutting-edge statistical technology still under development, and computers had only just been developed that could handle the thousands of computations required. Osgood and his colleagues proceeded to the task using Iliac I, a 5-ton "supercomputer" serving the entire campus of the University of Illinois, a computer that in fact was less powerful than some handheld calculators a half century later.

Results of the factor analyses, and of additional research applying semantic differentials to the topics of attitude change, psychopathology, and communications studies, were published in *The Measurement of Meaning* (Osgood, Suci, and Tannenbaum 1957). Chapter 2 of this book, focusing on the dimensionality of semantic space, applied several different methods of factor analysis to several different sets of data. The most elaborate analysis dealt with ratings of 20 concepts on 76 bipolar scales selected from *Roget's Thesaurus*, the judges being 100 undergraduates, each of whom was paid for three hours of participation. The researchers pooled individuals' ratings of different concepts to compute correlations among the scales. Factor analysis of the 76 × 76 correlation matrix yielded an Evaluation factor as the implicit dimension accounting for the most variation in ratings, a Potency factor as the second most important dimension, and an Activity factor as the third dimension. Other studies reported in the chapter analyzed judgments of adjective similarities, ratings of sonar signals on 50 bipolar scales, and ratings of representational artistic paintings on 40 bipolar scales. The authors summarized their work as follows (Osgood, Suci, and Tannenbaum 1957, pp. 72–73):

> When subjects differentiate the meanings of concepts, variance along certain scales (e.g., activity scales) may be quite independent of variation along other scales (e.g., evaluation). To put the matter yet another way, some of the things judged "good" may also be judged "strong" (e.g., HERO) but other things judged equally "good" may also be judged "weak" (e.g., PACIFIST). If meanings vary multidimensionally, then any adequate measuring instrument must encompass this fact. . . .
>
> In every instance in which a widely varied sample of concepts has been used, or the concept variable eliminated as in forced-choice among the scales, the same

> three factors have emerged in roughly the same order of magnitude. A pervasive *evaluative factor* in human judgment regularly appears first and accounts for approximately half to three-quarters of the extractable variance. Thus the *attitudinal* variable in human thinking, based as it is on the bedrock of rewards and punishments both achieved and anticipated, appears to be primary. ... The second dimension of the semantic space to appear is usually the *potency factor*, and this typically accounts for approximately half as much variance as the first factor—this is concerned with power and the things associated with it, size, weight, toughness, and the like. The third dimension, usually about equal to or a little smaller in magnitude than the second, is the *activity factor*—concerned with quickness, excitement, warmth, agitation, and the like.

Osgood and his colleagues extracted additional factors in their analyses, which they tried to name and interpret, but they found that these factors varied from one analysis to the next, unlike the first three extracted factors. Indeed, applying a criterion embraced later (Van de Geer 1993, p. 147), the displayed graph of factor importance (Osgood, Suci, and Tannenbaum 1957, Figure 4) suggests that only the first three factors were statistically trustworthy.

Reviews of *The Measurement of Meaning* (reprinted in Snider and Osgood 1969) were respectful, earnest, and penetrating, and they were scathing. A consensus among reviewers was that assessment of concepts on bipolar scales did not measure meaning, notwithstanding the book's title (although one reviewer thought the three dimensions might be a framework for differentiating meanings of adjectives). Linguist Uriel Weinreich (1958) excoriated the semantic differential as a means for assessing meanings, and he proposed that the technology served another function instead, measuring the affective associations of words, or their capacity for producing extralinguistic emotional reactions. Cross-cultural validations reported in the book were criticized for merely recovering peculiarities of English, transmitted to other languages via translation. Reviewers also criticized the book's statistical procedures, although these critiques ultimately were neutralized for the most part by later developments in statistics and by improved semantic differential studies.

A few years after the reviews appeared Osgood published a paper entitled "Studies on the Generality of Affective Meaning Systems" (Osgood 1962), which addressed most of the critiques leveled at *The Measurement of Meaning*. The very title of the paper was a concession to Weinreich's proposal that the semantic differential measures affect, and the paper went to considerable lengths to reconceptualize the semantic differential as a measure of affective association rather than as a measure of semantic meaning. As part of this revisionism, Osgood (1962, pp. 19–20) acknowledged that the three-dimensional structure emerging from factor analyses of bipolar scales corresponded to Wilhelm Wundt's affective dimensions: "The similarity of our factors to Wundt's tridimensional theory of *feeling*—pleasantness–unpleasantness, strain–relaxation, and excitement–quiescence—has been pointed out to me." He buttressed the affective nature of the Evaluation–

Potency–Activity dimensions by discussing similar dimensions emerging in studies of emotional expressions on people's faces. His general conclusion was that the appearance of corresponding dimensions in so many different realms occurred because affect was implicit throughout (Osgood 1962, p. 21).

> The highly generalized nature of the affective reaction system—the fact that it is independent of any particular sensory modality, yet participates with all of them—is at once the mathematical reason why *evaluation*, *potency*, and *activity* tend to appear as dominant factors and the psychological basis for synesthesia and metaphor. It is *because* such diverse sensory experiences as a *white* circle (rather than a black), a *straight* line (rather than crooked), a *rising melody* (rather than a falling one), a *sweet* taste (rather than a sour one), a *caressing* touch (rather than an irritating scratch) can all share a common affective meaning that one can easily and lawfully translate from one modality into another in synesthesia and metaphor. This is also the basis for the high interscale communalities which determine the nature and orientation of general factors.

Osgood began a massive cross-cultural project to address other complaints about *The Measurement of Meaning*, especially that the number of rated concepts in the factorial studies was too small and that the three-dimensional structure of adjectives in English had been reproduced in other languages through translations of English scales. Osgood's 1962 article sketched the plan of the cross-cultural project (Osgood 1962, pp. 14–15).

> An ideal design might be as follows: (1) We would use a sample of countries representing several different language families and diverse cultures; (2) we would obtain representative samples of polar qualifiers (e.g., adjectival opposites in English) independently in each language–culture community; (3) we would determine the factorial structure among these qualifier dimensions, (a) when simply related to each other directly and (b) when used as dimensions against which to rate a representative sample of concepts; (4) we would try to demonstrate that the factor structure remains essentially constant, (a) when bilinguals are compared in their two languages (to show that language code per se does not affect semantic factor structure) and (b) when monolinguals speaking different languages are compared (to show that cultural differences do not affect semantic factor structure).

Some preliminary results from the cross-cultural project were presented in Osgood's 1962 article. However, a comprehensive presentation of results took another 13 years to appear.

2.1.1 Cross-Cultural Project

Cross-Cultural Universals of Affective Meaning (Osgood, May, and Miron 1975) reported research conducted in 21 different culture–language venues: Afghanistan (Dari and Pashtu), Belgium (Flemish), Finland (Finnish), France (French), Greece (Greek), Hong Kong (Cantonese), India (Bengali, Hindi, and Kannada), Iran (Farsi), Italy (Italian), Japan (Japanese), Lebanon (Arabic), Mexico (Spanish), Netherlands (Dutch), Sweden (Swedish), Thailand (Thai), Turkey (Turkish), United States (white English), and Yugoslavia (Serbo-Croatian). At the time the book was being written, the authors reported that studies were also in progress in Costa Rica (Spanish), Germany (German), Hungary (Magyar), Malaysia (Bahasa Kebangsaan), Mexico (Tzeltal and Mayan), Poland (Polish), and the United States (black English); (listings of venues compiled from Osgood, May, and Miron 1975, Tables 4:15 and 5:2). Besides these culture–language venues, a computer printout sold in the University of Illinois Bookstore in 1978 (Heise 2009) contained data listings for Hebrew and Portuguese.

This mammoth cross-national project "was designed to test the hypothesis that, regardless of language or culture, human beings utilize the same qualifying (descriptive) framework in allocating the affective meanings of concepts" (Osgood, May, and Miron 1975, p. 6). This was to be accomplished by replicating cross-culturally the kinds of studies reported in the Osgood, Suci, and Tannenbaum (1957) book, without translating scales from American English to the other languages.

> To avoid the potential bias of translation, and the resultant ethnocentric bias, the procedures for selecting qualifiers that would eventually serve as the dimensions of judgment in SD [semantic differential] tasks had to be entirely intracultural; each language/culture group must determine its own descriptive scales. However, the overall methodology of these intraculturally independent samplings had to be standardized in order to make possible the intercultural comparisons required for testing the primary hypothesis of structural equivalence. (Osgood, May, and Miron 1975, p. 66)

Respondents (subjects) participating in the study at each venue were high school males.

> Since our purpose was to compare semantic systems *as a function of gross differences in both language and culture*, we wished to maximize equivalence of subjects (as representatives of their own languages and cultures), at least in the "tool-making" stages. At the preliminary Allerton conference [Monticello, Illinois; February, 1960] we decided to use young (14–18) male students in average high schools in urban settings. We wanted young high school–level people because they have full command of their native language and are generally integrated with their native culture but are less likely than college students, for example, to have been exposed to other cultures and languages. We decided to use only males, on the ground that education of females is much more variable (selective) than for males across the world. We wanted to use students in classroom situations because of the efficiency and inexpensiveness of collecting data with pencil-and-paper tests. The use of urban settings (usually) was dictated simply by the fact that our foreign colleagues were typically associated with universities, and these in turn were typically in urban areas. (Osgood, May, and Miron 1975, p. 20)

In order to derive bipolar scales indigenously, a multistep procedure was applied within each culture–language venue. First, substantive concepts that were recognizable in all of the cultures were used to elicit a large number of qualifiers used in that venue.

> The essence of the instructions was that the subjects were to place each of the 100 stimulus words in an appropriate linguistic frame and then complete the frame by supplying a single qualifier which, in their judgment, would fit the frame and the given noun. These instructions were modified to suit the grammatical requirements of each language. The particular test frames thus varied from language to language as the syntactical requirements for qualifier distribution varied. In American English the test frames were given as "The BUTTERFLY is—" and "The—BUTTERFLY," with subjects instructed to supply an appropriate adjective in the blank. (Osgood, May, and Miron 1975, p. 71)

The elicitation from 100 stimulus words was conducted with 100 indigenes, resulting in the collection of 10,000 qualifiers, many of which were duplicates. The total number of distinct qualifiers was reduced to 60 or so representative qualifiers via computerized procedures "designed to order qualifier types in terms of three criteria: *salience* (total frequency of usage across all substantives), *diversity* (number of different substantives with which used), and *independence* (lack of correlation with other qualifiers across substantives)" (Osgood, May, and Miron 1975, p. 78). The qualifiers chosen were then submitted to a panel of 10 sophisticated speakers of the indigenous language, who were instructed to give the opposite word for each qualifier, if one exists. After applying procedures to deal with qualifiers having multiple opposites and to deal with disagreements among panel members, 50 qualifier pairs were selected. Presuming that evaluations would predominate in pairs obtained this way, the field staff was instructed to add 10 additional pairs which they thought might be important locally even though the pairs did not get through the automatic data processing (Osgood, May, and Miron 1975, p. 101). The 60 pairs, converted to bipolar scales, later were pruned down to 50 scales for further application, on the basis of initial factor analyses within the given culture.

Opposite qualifiers were assigned to the two poles of bipolar scales, with seven checkmark positions between. These scales were used to rate the substantives that had been used to elicit qualifiers indigenously: "The original 100 substantives (nouns in English) are judged against these scales by 200 new subjects, also young high school–level males, in each language/culture community. The usages of scales are correlated and factor-analyzed, and the factor structures for different groups are compared" (Osgood, May, and Miron 1975, p. 112).

To illustrate, using the materials for white American English, the 100 substantives used to elicit qualifiers, and also used as stimuli for rating on the bipolar scales, consisted of the following: *anger*, *author*, *battle*, *belief*, *bird*, *book*, *bread*, *cat*, *chair*, *choice*, *cloud*, *color*, *courage*, *crime*, *cup*, *danger*, *death*, *defeat*, *doctor*, *dog*, *ear*, *egg*, *father*, *fear*, *fire*, *fish*, *food*, *freedom*, *friend*, *fruit*, *future*, *game*, *girl*, *guilt*, *hair*, *hand*, *head*, *heart*, *heat*, *hope*, *horse*, *house*, *hunger*, *husband*, *knot*, *knowledge*, *lake*, *laughter*, *life*, *love*, *luck*, *man*, *map*, *marriage*, *meat*, *money*, *moon*, *mother*, *music*, *need*, *noise*, *pain*, *peace*, *picture*, *pleasure*, *poison*, *policeman*, *power*, *progress*, *punishment*, *purpose*, *rain*, *respect*, *river*, *root*, *rope*, *seed*, *sleep*, *smoke*, *snake*, *star*, *stone*, *story*, *success*, *sun*, *sympathy*, *thief*, *thunder*, *tongue*, *tooth*, *tree*, *trust*, *truth*, *water*, *wealth*, *Wednesday*, *wind*, *window*, *woman*, and *work*.

The 50 bipolar scales in White American English were the following: *alive–dead*, *beautiful–ugly*, *big–little*, *burning–freezing*, *clean–dirty*, *deep–shallow*, *dry–wet*, *everlasting–momentary*, *faithful–unfaithful*, *fast–slow*, *fine–coarse*, *fresh–stale*, *full–empty*, *good–bad*, *happy–sad*, *heavenly–hellish*, *heavy–light*, *helpful–unhelpful*, *high–low*, *honest–dishonest*, *hot–cold*, *known–unknown*, *light–dark*, *long–short*, *many–few*, *mild–harsh*, *needed–unneeded*, *nice–awful*, *noisy–quiet*, *powerful–powerless*, *rich–poor*, *round–square*, *safe–dangerous*, *sane–mad*, *serious–funny*, *sharp–dull*, *shiny–dull*, *smart–dumb*, *smooth–rough*, *soft–hard*, *soft–loud*, *straight–crooked*, *strong–weak*, *sweet–sour*, *tender–tough*, *true–false*, *unbroken–broken*, *useful–useless*, *white–black*, *and young–old.*

For each language–culture venue, mean ratings of the substantives on the bipolar scales were factor analyzed in several ways: indigenously, biculturally with the scales of white American English, and panculturally with the scales from all other venues. The pancultural analysis was the crucial one for testing the hypothesis that human beings utilize the same framework of affective meaning, regardless of language or culture.

The pancultural analysis reported by Osgood, May, and Miron (1975) incorporated data from 21 language–culture communities. The data matrix consisted of 100 rows—one row for each of the rated substantives, and 1,050 columns—one column for each of the 50 scales in 21 different venues. Since the scales all had been used to rate the same things, the substantives, correlations could be computed among all 1,050 scales, and those correlations among scales could be factored. "This assumption [that the 100 substantives are constant in meaning across languages] is obviously contrary to fact in its extreme form, but . . . to the extent that our assumption of semantic constancy of concepts is *not* met, 'noise' is introduced, correlations are lowered, and the possibility of demonstrating meaningful pancultural factors is reduced" (Osgood, May, and Miron 1975, p. 135). Since N (the number of substantives) was 100, the correlation matrix could have no more than 100 factors, despite the large number of scales, but the investigators believed on the basis of accumulated experience that extracting just 10 factors would be adequate to represent essentially all of the common variance in the matrix. In any case, the crucial question was whether the first three factors would be recognizable as Evaluation, Potency, and Activity—with every culture contributing to the definition of those factors.

The results were unequivocal (Osgood, May, and Miron 1975, Table 4:15). The first factor (specifically, the first principal component of the correlation matrix) was Evaluation, with scales from every culture–language venue contributing to definition of the factor. The second factor was recognizable as Potency, and the third factor was Activity, and again scales from all venues contributed to these factors.

> The clearest evidence for shared semantic factors has come from the pancultural factorization. The first three factors were Evaluation, Potency, and Activity in that order. E was fully shared, as evidenced by high-loading scales of similar semantic flavor contributing to the factor from all communities; P was somewhat less shared, as evidenced by the scales of some communities having relatively low values on the common factor; A was more shared than P, even though its average loadings were lower. In the cases of both P and A, however, the same semantic flavor was apparent in all locations, many scales being translation-equivalent across communities as well. We feel that the theoretical purpose of this research has been achieved. We have been able to demonstrate conclusively that the three affective factors of meaning are truly pancultural—at least to the extent that our sample of some 21 language/culture communities is representative of all human societies. (Osgood, May, and Miron 1975, pp. 189–190)

Osgood, May, and Miron (1975, pp. 111–112) emphasized that their project purposively and conscientiously avoided any taint of ethnocentric projection.

> There was only one point at which translation from English into other languages was involved. This was translation of the list of 100 substantives, considered to be culture-common, to be used as stimuli in the elicitation of modes of qualifying. It should be noted that this one instance of translation was prior to any data collection, that the final list of 100 substantives has proved to be familiar to people of all cultures, and that our young male subjects were free to produce any qualifiers that happened to occur to them. The qualifier types obtained in each location—transliterated in some cases, but untranslated—were analyzed "blindly" by computer so as to generate a . . . list of qualifiers [that] was used to elicit opposites, and those for which familiar opposites existed were made into seven-step bipolar scales.

Thus, the discovery that the EPA structure of affect is pancultural was resistant to the kinds of criticisms directed by reviewers against the research reported in *The Measurement of Meaning* (Osgood, Suci, and Tannenbaum 1957).

2.1.2 Domains of Affect

Osgood fell ill a few years after the publication of *Cross-Cultural Universals of Affective Meaning* (Osgood, May, and Miron 1975), so that book was his final contribution to exploration of the affective dimensions and the measurement of sentiments. However, work on the three dimensions of affect continued. Scholl (2008) has provided a comprehensive review of the continuation studies, while also integrating research in areas that previously had acknowledged no basis in affect. Specifically, Scholl considered systems for classifying personalities, the general tenor of interpersonal behaviors, connotative word meanings, expressive displays of the body, and human emotions.

Scholl (2008, Section 1) reported that various studies in the latter half of the twentieth century found two basic dimensions underlying variations in interpersonal action and personality—love–hostility and dominance–submission, called the *interpersonal circumplex*; a third dimension, affect intensity, probably completes these interpersonal aspects of personality, and a similar third dimension of intensity/activity was found to describe relational behavior. Thus, personality and associated patterns of interpersonal behavior are best viewed within a three-dimensional system. Scholl noted that Osgood's research established the dimensions of Evaluation, Potency, and Activity as aspects of word meanings and verbal communication. In the case of nonverbal behavior, Scholl cited psychologist Albert Mehrabian as establishing three kinds of interpersonal bodily communications: those communicating sympathy by touching, proximity, leaning forward, mutual gaze, smiling, and lilted conversation; displays of relaxation indicating status and control, such as relaxed posture, asymmetrical arm and leg positioning, a sideways tilt, and placid hands; and displays indicating mobilization, such as quick motions and loud, rapid talking. Turning to classifications of feelings and emotions, Scholl found that a number of frameworks for representing these phenomena were based on a two-dimensional system of pleasure–displeasure and activation, yet studies additionally found that a third dimension of dominance–submission is required to distinguish unpleasant activated emotions such as anger versus fear.

Scholl's summed up his review of research regarding dimensions of interpersonal activity and communication as follows (2008, p. 10).

> We can conclude that probably three fundamental dimensions exist in all five areas of feelings, non-verbal and verbal communication, behavioral acts and personality traits, which parallel each other. The first is here called "*affiliation*" in line with a basic need for humans to affiliate and to get support for their survival; it entails sympathy/love (expressed for instance through acceptance, smile, closeness) versus antipathy/hostility (expressed through disgust, sarcasm, distance) as opposite poles, and it coincides with the more general dimension of evaluation (good–bad). A second dimension is here called "*power*" with dominance (expressed through anger, pride, or relaxation) versus submission (expressed through fear, awe, or tension) as opposite poles. The dimensional name is borrowed from the other well known interpersonal motive, the power motive, and it gets its fundamental significance by the fact that humans, like other social animals, form hierarchies and compete for dominance. Finally, a third dimension of "*activation*" or arousal was identified with the opposite poles of high arousal, active movements, loud voice, and fast speaking versus minimum arousal, passive, quiet, and slow behaviors. These three dimensions exhibit basic emotional as well as social qualities; therefore, we call them "*basic socio-emotional dimensions*."

Scholl's essay went on to examine how manifestations of the three dimensions influence one another across domains: For example, to what extent does a particular kind of personality get expressed in certain kinds of interpersonal activity, verbalizations, nonverbal displays, and emotions? Scholl found that research of this kind is sparse, notwithstanding its obvious significance in social psychology. He did express optimism however (Scholl 2008, Section 10): "There seems to be a clear promise in using these three socio-emotional dimensions for theory integration within psychological social psychology, between psychological and sociological social psychology, and between social psychology and other psychological disciplines [and even with] other human sciences like biology, sociology, and economics."

2.2 BIPOLAR SCALES

Between 1950 and 1975, Charles Osgood and his colleagues established that feelings and affective associations vary on dimensions of Evaluation, Potency, and Activity, in cultures around the world. *Evaluation* relates to aesthetics, morality, utility, hedonism, etc.; *Potency* relates to magnitude, strength, force, social power, expansiveness, etc.; and *Activity* relates to spontaneity, animation, speed, arousal, noise, etc. Osgood and his colleagues also developed semantic differential scales as a method for assessing the position of a feeling or impression in the three-dimensional affective space.

Ratings of concepts on semantic differential scales tapping Evaluation provide an efficient method for assessing the affective component of attitudes (Heise 1970b). More generally, ratings of concepts on semantic differential scales tapping all three dimensions provide an effective method of measuring sentiments, as conceived by William McDougall (1908, p. 437).

Semantic differential technology began with simple bipolar scales that could be produced on a typewriter for incorporation into printed or mimeographed questionnaires. The poles were characterized by adjectival opposites, and seven checkmark positions were distributed between the poles:

Good :_: :_: :_: :_: :_: :_: :_: Bad

Often, the middle position was labeled "neither, or equally," and the checkmark positions on either side were labeled with the adverbs "extremely," "quite," and "slightly" (Osgood, Suci, and Tannenbaum 1957, pp. 28–29). The labels provided an approximation to equal intervals between checkmark positions (see articles by Norman Cliff and Samuel Messick in Snider and Osgood 1969). Accordingly, the checkmark positions were coded numerically with numbers one to seven for data analyses, with the bad, weak, and inactive ends of scales receiving low numbers, and the good, strong, and active ends receiving high numerical codings. Alternatively, an arithmetically equivalent coding scheme could be used, ranging from –3 to +3, with the neutral position in the middle being coded zero.

I wrote two reports reviewing semantic differential technology in the late 1960s (Heise 1969c, 1970b). The more technical of these dealt with the EPA structure in factor analyses, with metric issues such as the assumptions of equal intervals and a zero point, and with sources of rating variance. Many of the controversies that I reviewed regarding the dimensionality of the affective space were neutralized by Osgood, May, and Miron's (1975) cross-cultural work, as discussed above. Issues regarding numerical coding of checkmark positions largely were nullified by later adoption of graphic bipolar scales anchored with multiple adjectives at each end. With regard to partitioning the variance of semantic differential ratings, I reviewed a number of studies and concluded that the variance of semantic differential ratings of a specific concept partitions as follows: "one-tenth due to subject–scale interaction, that is, due to differences between subjects in the use of scales, one-quarter due to bias and/or deviations of subjects' true scores from the population true scores, one-quarter due to momentary deviations of subjects from their own true scores and two-fifths due to random error" (Heise 1969c, p. 412). I revisit the issue of partitioning rating variance in this book, especially in Chapters 8 and 9.

My 1970 article, which offered detailed instructions on using semantic differentials in attitude research (advice that has been superseded in part by more recent technology), presented some benefits in using semantic differentials for the measurement of affective associations (Heise 1970b, pp. 248–250).

> *A generalized method*—An SD can be used as a generalized technique in the sense that a subject's attitude toward any object might be assessed by having the subject give ratings on the same set of SD scales. The SD offers the usual advantages of generalized attitude scales. (1) *Economy*. The same bipolar scales can be used to measure attitudes toward any object, so the costs of preparing a different scale for every object are eliminated. (2) *Instant Readiness*. An SD for measuring attitudes can be made up immediately for crash programs or for topical projects in social research like studies of disasters, riots or the appearance of new political figures. (3) *Cross-Concept Comparability*. Since attitudes toward various objects are all measured on the same scales, there is the potential for comparing different attitudes. The major problem in using a single set of SD scales as a generalized attitude scale is the matter of scale relevancy or, more generally, of concept-scale interactions. A single set of scales used for all objects would provide relatively insensitive measurements for some. This may not be objectionable (for example, in exploratory work or in research involving a large number of attitude objects) where a set of scales like those from the pan-cultural factor analyses can be used as a rough and ready instrument for general attitude measurement. Where, however, sensitivity is necessary, it probably is desirable to use the SDs developed for the particular content areas of interest.

> *Standard metric*—One of the unique features of the SD is that attitudes toward a vast array of objects can be measured in terms of basically the same metric on the three EPA dimensions. Thus, all of the objects can be positioned in a single attitudinal space. This feature of the SD has yielded developments and insights that would have been difficult or impossible to obtain using attitude scales in which the metric changes for each object considered. . . .
>
> *Cross-cultural comparisons*—With cross-cultural validation and extension of SD measurements, the way is opened for cross-cultural comparisons of attitudes. This work has barely begun, but already results suggest that while major cultural variations do exist in attitudes toward various objects, there are also some striking uniformities. GIRLS, LOVE, and MARRIAGE, for example, seem to be positively evaluated in several cultures. Much of the variation in attitudes is ecologically determined by the nature of the objects and, even cross-culturally, there is less variation in attitudes than one might expect.

The fact that semantic differentials can serve as a generalized method of measuring sentiments has fostered the creation of repositories of cultural sentiments, recording subjective cultures in several different nations. This research, and the use of the repositories to make cross-cultural comparisons, are reviewed in Chapter 3.

Semantic differentials use a single measurement metric for the three dimensions of sentiments: Evaluation, Potency, and Activity. Thereby, sentiments regarding social identities, social behaviors, social settings, and individual attributes are represented on the same scale. This standardization enabled the development of empirical equations describing impression formation processes in humans. The equations are obtained by presenting vignettes describing various interpersonal events and regressing measurements of emergent feelings about the components of the events on measurements of pre-event sentiments regarding the components. (This research is reviewed in Chapter 4.)

Quantifications of impression formation processes in turn fostered development of a mathematized theory of social interaction: affect control theory (Heise 1979, 2007; MacKinnon 1994; Smith-Lovin and Heise 1988). Affect control theory converts empirically derived impression formation equations into proaction equations for predicting interpersonal actions, under the assumption that people try to create impressions supporting their sentiments. The standardized metric of the semantic differential was fundamental in developing affect control theory because it enabled multidimensional analysis of descriptions of social events comprised of identity nouns, setting nouns, behavior verbs, and emotion and trait adjectives. Whereas "innumerable theories offer explanations for how subsets of these elements [identities, settings, actions, and emotions] are related in particular contexts, [affect control theory] offers a general explanation for the entire set of relationships . . . an astounding achievement" (Clore and Pappas 2007, p. 333). My 2007 book presents affect control theory in detail. Part 1 provides a plain-language exposition of the theory, along with numerous interpretive analyses of everyday situations; Part 2 presents the mathematical derivations that define sentiment-confirming behavior, labeling, attribution, and emotion; and Part 3 describes the research program associated with the theory and the computer simulation software. Affect control theory's simulation software in conjunction with repositories of measured cultural sentiments fosters explorations of how a group's subjective culture translates into many forms of social action.

In the 1970s and thereafter, semantic differential scales transmuted from their original form as bipolar scales with a single adjective at either end of seven checkmark positions into new forms that addressed recognized problems with the measurement system and that incorporated new technology.

2.2.1 Concept–Scale Interaction

A phenomenon called concept–scale interaction loomed large in many methodological studies of the semantic differential (e.g., Heise 1969c; Osgood, May, and Miron 1975; Osgood, Suci, and Tannenbaum 1957).

> Although such scales as *sweet–sour*, *merciful–cruel*, *light–gloomy*, *ambrosial–poisonous*, and *friendly–repelling* are denotatively quite distinct, they are used affectively in the same way and represent a common "evaluative" factor. Nevertheless, the fact that the scales *do* have different denotation despite their common affective components means that they may be used literally with certain relevant concepts, producing what we call "concept-scale interaction." Thus for a few concepts (e.g., LEMON, COCA-COLA, SUGAR) the English "evaluative scale" *sweet–sour* will be used denotatively. (Osgood, May, and Miron 1975, p. 33)

Concept–scale interaction arises from two different phenomena in measuring attitudes and sentiments with semantic differentials. To illustrate the first, which I'll call *designative*, the concept *lemon* would probably be rated as quite sour by most respondents, giving it an evaluation rating of –2.0; however, its actual evaluation (based on the scales *good–bad* and *kind–cruel*) is +1.0 (Heise 1978, p. 129). Distortion from ratings on a designative scale continues even after averaging them with ratings on other scales: for example, if ratings of lemon on the three scales *good–bad*, *kind–cruel*, and *sweet–sour* were averaged, the result would be 0.0 instead of a positive value. The second relevant phenomenon was identified by Daniel Kahneman (1963) as a tendency for some respondents to exaggerate their ratings while other respondents attenuate their ratings. This leads to correlations among scales when used to rate concepts that are extreme on those scales (Kahneman 1963, p. 562): "If, for instance, the true scores of *Adlai Stevenson* indicate that he is both PLEASURABLE and MASCULINE, the correlation between the two scales [pleasurable–painful and masculine–feminine] will be positive. For the concept *My Mother*, which is PLEASURABLE and FEMININE, the sign of the correlations will be reversed." Kahneman was able to explain all but obvious cases of designative concept–scale interaction with the exaggeration factor. The exaggeration factor is considered in detail in this book, except that it is interpreted here as an effect arising from some respondents' limited inculcation into the culture being assessed by a survey (see Chapter 5).

Averaging ratings on a large number of scales associated with a particular affective dimension offers one approach to minimizing the errors created by designative concept–scale interaction. For example, if *sweet–sour* were one of 10 evaluative scales, and the average rating on all the others were 1.0, the average including *sweet–sour* would be 0.7, which is fairly close to the true value. However, expanding the number of scales used to measure sentiments depletes research resources rapidly: 10 scales for each dimension—Evaluation, Potency, and Activity—requires respondents to make 30 ratings for each concept, so a respondent might have time to rate only 10 to 15 concepts at a sitting.

My solution to this conundrum as I began amassing repositories of cultural sentiments in the 1970s was to anchor each end of a single bipolar scale for each dimension with multiple adjectives, the sets of adjectives being assembled from the scales that defined the given dimension in factor analyses. For example, the four recommended scales for measuring Evaluation in American English are *nice–awful*, *good–bad*, *sweet–sour*, and *helpful–unhelpful* (Osgood, May, and Miron 1975, Table 4:18). I created a measuring instrument with a single scale to measure Evaluation, anchored by the cluster of adjectives *good, helpful, nice, sweet* on one side and the adjectives *bad, unhelpful, awful, sour* on the other side (Heise 1978, Figure 4.1). Thereby, the designative impact of any specific pair of adjectives should be subordinated to the overall metaphoric sense of the adjectives as a set of evaluators. Furthermore, the economic gain was substantial; instead of each respondent being able to rate just 10 concepts with 30 scales at a sitting, the respondent could rate 100 concepts with three scales.

2.2.2 Computerized Graphic Scales

My early projects with semantic differentials (Heise 1965, 1969a, 1970a) revealed another methodological problem: The seven rating positions in early semantic differentials sometimes were too few to record the full range of respondents' feelings. I converted to nine-step scales for this reason.

> The usual seven-point semantic differential scale is subject to ceiling and floor effects because of its limited range; for example, both Friend and God might be rated identically as "extremely good," "extremely" being the adverb that defines the outermost position on a scale. Ordinarily ceiling and floor measurement errors are attenuated by averaging over several scales which do not have identical ceiling–floor characteristics. Since only a single scale is being employed for each dimension here, however, scale bounds were extended by one position at each end, giving a nine-point scale with positions labeled "infinitely," "extremely,"

"quite," and "slightly" (used on both sides of a scale) plus a middle position labeled "neither or neutral." Respondents were instructed to use the "infinitely" category when they felt that a stimulus had as much of a given trait (goodness, badness, power, liveliness, etc.) as anything could. (Heise 1978, p. 64)

Ratings of 650 social identities and 600 social behaviors on the nine-step scales were analyzed with successive-intervals scaling to determine the underlying metric guiding respondents' use of the nine checkmark positions (Heise 1978, pp. 70–79). These metric analyses revealed some deviations from the equal-intervals assumption typically employed with semantic differentials; successive-intervals scaling also revealed that scale metrics varied somewhat with the kinds of stimuli being rated (nouns or verbs), the orientation of the scales (good or powerful or active on the left versus on the right), and the dimension being measured. Thus, derived metrics were used to code respondents' choices for subsequent analyses with these kinds of scales. Interestingly, the "infinitely" checkmark position was just 1.0 to 1.5 as far from the "extremely" checkmark position as the "extremely" position was from the "quite" position, a finding that was confirmed in a later analysis of an even larger dataset (Smith-Lovin 1987b, Figure 1).

I developed a computerized rating system, program *Attitude*, which was used in collecting data on sentiments in the 1980s and 1990s (Heise 1982b). The program presented a stimulus to be rated at the top of a computer's screen and a single semantic differential scale at the middle of the screen. The scale was represented in graphic form as a line, and adjective clusters appeared at either end of the line to define the dimension being rated: for Evaluation—*bad, awful* versus *good, nice*; for Potency—*little, powerless* versus *big*, powerful; and for Activity—*slow, old, quiet* versus *fast, young, noisy*. The adjectives changed after each rating and after each change in stimulus. Quantifiers "infinitely," "extremely," "quite," and "slightly" appeared below the line to suggest segmentation, and the word "neither" indicated the midpoint of the scale. A checkmark appeared above the line, initially at the midpoint. The respondent moved the checkmark to the left or right with arrow keys on the computer keyboard, up to 19 positions to the left and 19 to the right, plus the neutral position, a total of 39 increments in the scale. A respondent's final positioning of the checkmark was coded in terms of the magnitude of the distance of the checkmark from the scale midpoint, converted to numerical scale values from –4.75 to +4.75. Coding of distances on the graphic scale circumvented the problem of deriving a quantitative metric for qualitative steps in a scale. A similar procedure for sentiment measurement had been used successfully by psychologist Harry Gollob since the 1960s.[1]

[1]Gollob's printed questionnaires presented evaluation scales as lines with bipolar verbal anchors; numerical quantifiers on the line ranged out from the zero midpoint: minus on one side and plus on the other side. Respondents were instructed to record their judgment as a mark on the line; the marks were coded by measuring their positions with a ruler (Harry Gollob, personal communication, c. 1979). Some of Gollob's work is discussed in Chapter 4 of this book.

Development of the World Wide Web in the 1990s raised new opportunities. I realized that adapting the data collection program for the Web would make it easier to access, allowing data to be collected easily from people anywhere in the world. So I rewrote the computerized rating system as a Java applet called *Surveyor* in 1997. The applet performs computer-assisted self-administered interviews and returns the acquired data via the Internet. This is the tool currently used for collecting sentiment data, and it was used specifically to collect the empirical data introduced in Chapter 3 and used throughout much of this book. Therefore, I describe the instrument in detail later in this chapter.

2.2.3 Iconic Scales

The cross-cultural project presented by Osgood, May, and Miron (1975) developed indigenous bipolar scales to avoid the ethnocentricity of translating scales from English. However, in their final chapter the authors considered whether there might be a more direct procedure for obtaining scales for use in different cultures. Perhaps, they speculated, instead of anchoring the ends of bipolar scales with words, the scales could be anchored with images, and the same images would work for humans in any culture. After all, semantic differentials developed from research on synesthesia, the phenomenon in which sensations in one sense domain cause sensations in a different sense domain, and synesthesia is mediated by the affective associations of sensations (Osgood, May, and Miron 1975, pp. 396–400). So perhaps specific kinds of images would function the same as verbal metaphors used to define scales.

Osgood, May, and Miron (1975, pp. 381–384) reported a study conducted in the United States, Finland, Germany, India, and Japan in which 50 concepts were rated in each culture on indigenous EPA scales from the pancultural study and also on 64 seven-step scales defined with a variety of oppositional pictograms (e.g., a curvy line versus a jagged line). The results were mixed.

> (1) Using the pancultural SD scales as markers of E, P, and A, a multicultural factor analysis yielded clear evidence for a universal E factor (defined by shared sets of symmetrical vs. nonsymmetrical forms). There was reasonably good evidence for P and A for [Finland, Germany, and Japan] but not for [the United States and India], and the high-loading pictorial scales varied much more than for E. (2) There was evidence for "denotative contamination" for P (e.g., *angular* vs. *rounded* pictograms correlated highly with concepts like ANGER, CHAIR, and POLICEMAN vs. CLOUD, SMOKE, and SNAKE. (3) When the data for Abstract vs. Concrete verbal concepts were analyzed separately, the correspondence between verbal and visual E-P-A scales proved to be much higher for the low-imagery Abstract concepts—as might be expected from the "contamination" effects above. (Osgood, May, and Miron 1975, p. 381)

Osgood, May, and Miron observed a number of flaws in the design of the study, and the pictograms also failed to reflect the diversity of images that had been scored on EPA by Elliot and Tannenbaum (1963), especially with respect to the Potency and Activity dimensions. So the study may not have provided an adequate test regarding the use of pictograms for assessing affective associations in cross-cultural contexts.

Peter Raynolds and his associates (McCulloch and Raynolds 2007; Raynolds and Raynolds 1989; Raynolds, Sakamoto, and Raynolds 1988; Raynolds, Sakamoto, and Saxe 1981) used inkblots from the Rorschach and Holtzman projective tests, plus additional inkblots specially made, to obtain abstract images that contrast on affective dimensions. They presented pairs of images briefly enough to prevent any kind of rational analysis, and respondents chose the image that seemed more like the stimulus concept. Results correlated with semantic differential ratings for topics where respondents have no internal conflicts, yet the method gets at nonconscious feelings which may be different than normative semantic differential results in sensitive areas, to the point of revealing personal information beyond respondents' usual awareness. A cross-cultural study in the United States and Japan showed that most image pairs have the same significance in these two societies, but forms within some of the abstract images "might resemble kanji characters, and so influence Japanese but not American responses" (Raynolds, Sakamoto, and Raynolds 1988, pp. 400–401).

Peter Lang devised cartoon-based scales for measuring the dimensions of pleasure, arousal, and dominance. His Self-Assessment Manikin (SAM) approach defines the pleasure scale with a smiling, happy figure at one end, transforming to a frowning, unhappy figure at the other end. For the arousal dimension, the figure at one end of the scale is an excited wide-eyed figure and a relaxed, sleepy figure is at the other end. The dominance dimension is represented by a large figure at one end of the scale and a tiny figure at the other end (Bradley and Lang 1994, pp. 50–51).

The pictorial constitution of SAM scales explicitly was intended to provide a cross-cultural measurement methodology (Bradley and Lang 1994, p. 50), but no cross-cultural validation was offered, and indeed, presentations of the instrument made no reference to the major cross-cultural study of the semantic differential (Osgood, May, and Miron 1975), although *The Measurement of Meaning* (Osgood, Suci, and Tannenbaum 1957) was a key source. Furthermore, instructions to respondents undercut cross-cultural adaptability by identifying the meanings of the block cartoon figures on each scale with words from corresponding semantic differential scales (Bradley and Lang 1994, p. 53).

SAM ratings of 21 pictures correlated highly with semantic differential ratings of the same pictures on the pleasure and arousal dimensions. However, SAM dominance ratings correlated mainly with pleasure (Bradley and Lang 1994, Table 2), possibly because the dominance scale actually assesses pride (expansive) versus shame (shrinking). Mean ratings of 1,034 English words by undergraduate males and by undergraduate females are available (Bradley and Lang 1999).

SAM's accuracy and cross-cultural adaptability might be improved by focusing specifically on the face, which is humans' primary instrument for communicating emotion cross-culturally. Drawing on decades-long research of Paul Ekman (2004), I devised computer-based procedures for translating an emotion's Evaluation–Potency–Activity profile into a sketchlike representation of the corresponding facial expression. The first step was condensing Ekman's findings enough to apply to simple drawings (Heise 1982a, p. 31).

> Emotional messages are constructed on the face by the shape of the mouth, eyes, and eyebrows (and sometimes the nose, cheeks, and forehead as well). Each of these features has a limited number of major shapes produced by the action of certain facial muscles. Whether a group of muscles is tugging gently or straining hard may suggest the intensity of feeling, but the real information is in the fact that certain muscles are operative, producing the characteristic shape for that muscle group.
>
> The brows have four major shapes other than a neutral relaxed position. They may be curved upward (as in surprise), flattened and raised (as in fear), flattened and lowered (as in sadness), or pulled down and inward (as in anger).
>
> The opened eyes have six major shapes: neutral, wide open (as in surprise), raised lower lids (as in disgust), raised and tensed lower lids (as in fear), squinting (as in anger), and upper lids drooping and sloped (as in sadness).
>
> Major shapes of the mouth, aside from neutral, are: dropped open (as in surprise), corners pulled horizontal (as in fear), lips pressed tight (as in anger), squared outthrust lips baring teeth (as in anger), upper lip pulled up (as in disgust), corners down (as in sadness), corners raised (as in happiness, with extra stretching for smiles, grins, or laughs).
>
> The end of the nose may be normal or raised by pressure from the upper lip; the upper nose may be normal or crinkled. Cheeks may be normal or raised during laughter. The forehead may be normal or wrinkled by pressures from the eyebrows.
>
> Variations in one feature combine with variations in another feature; for example, any eyebrow formation can occur with any mouth shape. But not quite every combination of features is possible. For example, the mouth isn't disgusted alone; "disgusted" mouth occurs with nose raised.
>
> Expressions for the primary emotions are universal. Surprise combines arched eyebrows with wide open eyes and a dropped open mouth. Fear shows in raised and flattened eyebrows, raised and tensed lower eyelids, along with side stretched lips. Disgust involves raised lower eyelids, and the upper lip curled up so to raise the nose; the upper nose may be crinkled. In anger the brows pull down and inward, the eyes squint, and the lips either are pressed tight or squared into a snarl. Happiness is revealed in upturned corners of the mouth; laughing also raises the cheeks which in turn may push the lower eyelids up.

My 1982 article described a computer program that allowed a user to control the various facial features from the computer's keyboard and thereby to generate innumerable emotional expressions in a face drawn on the computer screen. Later I linked changes in facial features to EPA profiles of emotions so that a program for simulating social interaction could show the emotional expressions of interactants (Heise 2004). The basic rules for converting EPA profiles of emotions into emotional expressions on the face are as follows: open eyes with positive A; arch up brow with positive E; raise brow with negative P, lower brow with positive P; move mouth higher with positive P, and move upper lip higher with positive P; drop lower lip and narrow mouth with positive A; curve lips up with positive E, down with negative E.

It would be straightforward to create a computer program with which people could create facial expressions corresponding to their own feelings in response to stimuli, and then have the computer convert the expression that the respondent draws to an implied EPA profile. Left–right motions with a computer mouse could instigate changes in facial expression corresponding to calmness, happiness, joy, excitement, fear, and depression; up–down motions could increase or decrease dominance in order to register additional feelings, such as rage, annoyance, and disgust (see the spiral model of emotions in Figure 7.1 in Francis and Heise 2008). Since Ekman showed that emotional expressions communicate feelings cross-culturally (Ekman 1971), such a system has a reasonable chance of having cross-cultural validity as a sentiment-measurement device, although that validity would have to be ascertained empirically, of course.

2.2.4 Creating New Adjective Scales

The cross-cultural project reviewed above yielded sets of scales with pancultural communality in numerous language communities. In particular, Osgood, May, and Miron (1975, Table 4:18) listed four adjective pairs that they recommended for measuring each dimension in 21 cultures. Table 2.1 shows their selections and the numerical data that they included with their selections. Under "Pancultural E" in Table 2.1 are lists of four scales per culture for measuring the Evaluation dimension, where the Evaluation dimension has been defined in a pancultural factor analysis of 21 language communities. The numbers following each adjective pair are the absolute values of the factor loadings on the first three factors in the pancultural analysis. Similarly, recommended scales for measuring Potency, followed by absolute values of each scale's factor loadings, are listed under "Pancultural P"; and recommended scales for measuring Activity, with absolute values of factor loadings, are listed under "Pancultural A." Adjective pairs are given in English, having been translated from the various indigenous languages.

I supplemented the Osgood, May, and Miron selections with scale selections for three additional language communities—black American English, Brazilian Portuguese, and Malay—that were available in the computer printout sold at the University of Illinois Bookstore in the 1970s (Heise 2009). The scales recommended for these cultures (i.e., the highest loading scales on each

TABLE 2.1 Scales for Measuring Evaluation, Potency, and Activity (EPA) in 24 Cultures, with Absolute Values of Loadings from Pancultural Factor Analysis

Pancultural E				Pancultural P				Pancultural A			
	E	P	A		E	P	A		E	P	A
ENGLISH											
nice–awful	.94	.15	.08	powerful–powerless	.22	.68	.08	fast–slow	.14	.20	.61
good–bad	.92	.09	.12	strong–weak	.11	.57	.13	young–old	.32	.39	.45
sweet–sour	.90	.14	.00	deep–shallow	.09	.57	.11	noisy–quiet	.30	.34	.40
helpful–unhelpful	.89	.09	.09	big–little	.01	.68	.29	alive–dead	.52	.12	.55
FRENCH											
pleasant–unpleasant	.90	.13	.01	strong–weak	.08	.59	.00	lively–languid	.20	.24	.61
good–bad	.89	.05	.04	huge–tiny	.01	.57	.13	fast–slow	.17	.23	.57
nice–wicked	.09	.06	.09	low-pitched–high-pitched	.11	.43	.04	living–dead	.48	.06	.56
marvelous–awful	.86	.02	.00	big–little	.23	.68	.14	young–old	.55	.17	.42
FLEMISH											
good–bad	.91	.10	.08	strong–weak	.20	.58	.15	quick–slow	.16	.16	.69
magnificent–horrible	.89	.08	.04	big–small	.01	.57	.20	active–passive	.24	.34	.65
beautiful–ugly	.88	.09	.02	long–short	.09	.42	.12	sanguine–phlegmatic	.27	.12	.42
agreeable–unagreeable	.88	.13	.00	deep–shallow	.02	.50	.05	shrewd–naive	.33	.27	.40
DUTCH											
happy–unhappy	.91	.06	.16	big–little	.02	.57	.15	fast–slow	.10	.28	.71
pleasant–unpleasant	.91	.12	.10	heavy–light	.31	.55	.30	active–passive	.04	.35	.72
good–bad	.90	.09	.01	strong–weak	.10	.54	.00	excitable–calm	.32	.29	.45
pretty–not pretty	.87	.08	.16	hard–soft	.35	.46	.31	fascinating–dull	.29	.35	.49
SWEDISH											
good–bad	.86	.10	.02	high–low	.01	.50	.25	sanguine–not sanguine	.19	.10	.66
nice–nasty	.84	.17	.04	strong–weak	.03	.47	.17	quick–slow	.14	.16	.63
right–wrong	.82	.04	.13	long–short	.02	.45	.26	lively–lazy	.20	.14	.62
kind–evil	.82	.13	.13	difficult–easy	.30	.51	.01	active–passive	.11	.22	.54

TABLE 2.1 **(*Continued*)**

Pancultural E				Pancultural P				Pancultural A			
	E	P	A		E	P	A		E	P	A
FINNISH											
nice–not nice	.88	.12	.08	large–small	.09	.60	.26	fast–slow	.02	.11	.67
pleasant–unpleasant	.87	.12	.04	sturdy–delicate	.19	.59	.38	flexible–rigid	.22	.03	.67
good–bad	.86	.10	.13	strong–weak	.10	.39	.14	agile–clumsy	.01	.13	.64
light–gloomy	.81	.08	.19	heavy–light	.19	.51	.33	lively–subdued	.40	.08	.53
SPANISH (MEXICAN)											
admirable–despicable	.93	.02	.08	strong–weak	.06	.54	.10	active–passive	.01	.23	.56
friendly–repelling	.92	.01	.07	big–small	.23	.58	.23	fast–slow	.21	.09	.44
agreeable–disagreeable	.93	.13	.11	giant–dwarf	.29	.60	.18	young–old	.40	.22	.46
good–bad	.92	.10	.09	major–minor	.24	.55	.04	soft–hard	.34	.30	.37
ITALIAN											
valuable–worthless	.93	.04	.05	big–small	.09	.66	.19	fast–slow	.14	.15	.66
beautiful–ugly	.92	.10	.04	strong–weak	.12	.55	.02	mortal–immortal	.09	.08	.48
good–bad	.92	.14	.01	wide–narrow	.28	.54	.20	young–old	.31	.11	.48
desirable–undesirable	.92	.12	.08	high, tall–short	.27	.49	.06	hot–cold	.51	.02	.39
SERBO-CROATIAN											
pleasant–unpleasant	.93	.01	.04	big–little	.25	.72	.08	alive–dead	.37	.23	.63
good–bad	.92	.01	.02	bulky–tiny	.12	.67	.21	fast–slow	.07	.34	.54
lovable–disgusting	.91	.00	.02	strong–weak	.09	.67	.06	young–old	.48	.17	.45
bearable–unbearable	.88	.01	.05	unlimited–limited	.07	.42	.16	soft–hard	.30	.35	.41
GREEK											
superb–awful	.93	.03	.07	big–small	.08	.60	.25	quick–slow	.14	.11	.55
good–bad	.91	.04	.02	strong–weak	.18	.60	.19	young–old	.23	.04	.52
friendly–hostile	.88	.01	.06	difficult–easy	.04	.39	.04	curious–indifferent	.06	.06	.35
useful–useless	.85	.07	.15	brave–cowardly	.36	.46	.25	active–passive	.33	.20	.39

TURKISH											
beautiful–ugly	.91	.11	.02	big–little	.11	.67	.13	fast–slow	.02	.32	.50
good–bad	.90	.12	.13	heavy–light	.24	.58	.26	alive–not alive	.41	.01	.49
tasteful–tasteless	.90	.15	.02	strong–weak	.19	.43	.10	young–old	.36	.20	.48
pleasant–unpleasant	.90	.16	.04	deep–shallow	.00	.42	.01	rough–calm	.29	.18	.37
ARABIC											
beautiful–ugly	.90	.06	.13	large–small	.37	.51	.29	fast–slow	.05	.27	.35
good–bad	.90	.04	.17	long–short	.19	.40	.33	hot–cold	.21	.28	.31
safe–dangerous	.90	.09	.15	strong–weak	.43	.42	.14	emotional–logical	.12	.30	.28
vivifying–fatal	.89	.00	.12	open–closed	.31	.33	.03	thin–thick	.49	.15	.30
FARSI (IRANIAN)											
good–bad	.92	.09	.09	heavy–light	.17	.60	.20	active–inactive	.31	.34	.53
worthwhile–worthless	.89	.05	.08	thick–thin	.02	.47	.31	fast, sharp–slow, dull	.08	.38	.41
auspicious–ominous	.88	.00	.06	stout–slim	.19	.43	.12	exciting–spiritless	.47	.30	.52
best–worst	.88	.05	.13	strong–weak	.31	.42	.02	warm–cold	.53	.17	.29
DARI (AFGHAN)											
good–bad	.88	.06	.06	military–civil	.27	.45	.11	nimble–lazy	.33	.17	.52
auspicious–inauspicious	.86	.09	.08	great–small	.49	.55	.14	sharp–dull	.31	.19	.41
safe–dangerous	.84	.00	.03	long–short	.24	.30	.11	soft–hard	.30	.26	.41
lovely–ugly	.83	.13	.01	exaggerated–sparse	.26	.28	.05	fast–slow	.24	.15	.33
PASHTU (AFGHAN)											
loving–hating	.86	.01	.07	big–small	.21	.54	.17	alive–dead	.19	.30	.61
attractive–unattractive	.88	.05	.10	brave–cowardly	.16	.47	.16	fast–slowly	.08	.25	.48
tasty–tasteless	.83	.00	.01	strong–weak	.14	.35	.09	soft–hard	.04	.14	.46
good–bad	.88	.02	.19	high–low	.32	.42	.16	awake–asleep	.14	.38	.44
HINDI											
glad–not glad	.83	.03	.04	difficult–easy	.01	.44	.05	gay–sober	.08	.14	.47
good–bad	.83	.01	.14	heavy–light	.09	.46	.25	narrow–wide	.04	.02	.29
ambrosial–poisonous	.81	.01	.09	brave–cowardly	.54	.47	.12	loquacious–taciturn	.04	.11	.30
superior–inferior	.80	.02	.19	fast–slow	.05	.47	.32	thin–thick	.02	.12	.36

TABLE 2.1 (*Continued*)

Pancultural E				Pancultural P				Pancultural A			
	E	P	A		E	P	A		E	P	A
BENGALI											
beautiful–ugly	.93	.06	.03	huge–minute	.28	.62	.20	fast–slow	.40	.27	.43
lovely–repulsive	.93	.04	.06	powerful–powerless	.09	.60	.11	industrious–lazy	.24	.44	.43
kind–cruel	.91	.02	.03	big–little	.28	.54	.23	alive–dead	.50	.27	.48
superior–inferior	.91	.01	.12	strong–weak	.16	.55	.00	thin–thick	.06	.08	.30
KANNADA											
merciful–cruel	.78	.10	.01	big–small	.04	.41	.14	active–dull	.36	.24	.50
good–bad	.76	.04	.09	wonderful–ordinary	.18	.45	.15	loose–tight	.14	.07	.36
calm–frightful	.74	.19	.01	great–little	.01	.34	.18	fast–slow	.22	.15	.33
delicate–rough	.75	.16	.04	huge–small	.28	.41	.23	unstable–stable	.35	.07	.34
THAI											
useful–harmful	.88	.04	.09	heavy–light	.08	.50	.18	quick–inert	.02	.24	.56
pure–impure	.86	.03	.03	deep–shallow	.18	.49	.11	fast–slow	.25	.14	.44
fragrant–foul	.86	.01	.03	loud–soft	.11	.43	.20	thin–thick	.06	.36	.40
comfortable–uncomfortable	.86	.09	.09	hard–soft	.18	.42	.34	little–much	.23	.14	.25
CANTONESE											
lovable–hateable	.92	.09	.05	big–little	.18	.75	.01	agile–clumsy	.33	.17	.68
good–poor	.91	.01	.08	tall, big–short, small	.12	.76	.19	fast–slow	.18	.18	.54
good–bad	.92	.07	.13	strong–weak	.02	.72	.18	red–green	.08	.24	.43
respectable–despicable	.90	.03	.07	deep–shallow	.02	.63	.05	alive–dead	.53	.16	.49

JAPANESE											
good–bad	.93	.01	.18	heavy–light	.17	.66	.29	fast–slow	.05	.13	.42
comfortable–uncomfortable	.92	.04	.12	big–little	.25	.63	.21	vivid–subdued	.05	.06	.42
happy–sad	.91	.05	.02	difficult–easy	.25	.59	.02	noisy–quiet	.45	.20	.48
pleasant–unpleasant	.91	.07	.16	rare–common	.07	.50	.02	cheerful–lonely	.39	.03	.37
				From Atlas Printout							
AFRICAN-AMERICAN ENGLISH											
all right–mad	.80	.04	.17	large–small	.10	.63	.05	fast–slow	.18	.08	.36
good–foul	.77	.15	.12	big–small	.11	.58	.04	active–passive	.24	.14	.34
clean–nasty	.76	.08	.23	little–big	.23	.49	.04	beat–straight	.62	.18	.21
together–wrong	.75	.13	.20	thick–thin	.14	.28	.10	jive–straight	.56	.29	.19
PORTUGUESE (BRAZIL)											
desirable–undesirable	.93	.03	.02	big–small	.14	.72	.07	fast–slow	.13	.19	.69
good–bad	.92	.02	.02	heavy–light	.28	.60	.02	active–passive	.13	.14	.54
marvelous–horrible	.89	.08	.03	wide–narrow	.21	.60	.24	mortal–immortal	.09	.33	.50
comfortable–uncomfortable	.85	.07	.05	high–low	.17	.54	.00	noisy–silent	.26	.05	.47
MALAY											
excellent–bad	.93	.04	.04	long–short	.02	.72	.18	quick–slow	.04	.03	.73
affectionate–hating	.91	.08	.04	big–little	.20	.72	.05	fast–slow	.29	.10	.67
good–evil	.91	.02	.11	broad–narrow	.42	.58	.06	sharp–dull	.13	.11	.56
fitting–not fitting	.90	.06	.06	thick–thin	.14	.58	.03	brave–afraid	.37	.24	.48

Source: Based on Osgood, May and Miron's (1975) Table 4:18 (copyright © 1975 by the Board of Trustees of the University of Illinois Press; used with permission of the University of Illinois Press), with supplements from *Atlas* printout (Heise 2009).

factor) are given at the end of Table 2.1. Factor loadings in these cases are from a pancultural analysis of 24 cultures, and the absolute values of the loadings are displayed to maintain continuity with the upper part of the table.

A sentiment-measuring instrument can be created for any of the language communities listed in Table 2.1 by translating the given adjectives back into the target language and using the results as anchors for bipolar scales. Factor loadings given in Table 2.1 that indicate Evaluation scales created this way will provide reliable measurements with little contamination from the other dimensions. (That is, the Evaluation scales have high loadings on the pancultural Evaluation factor and low loadings on the other pancultural factors.) In most cultures Potency scales created from Table 2.1 will also be reliable and pure. Factor loadings indicate that Activity scales created this way will have just moderate reliability and some contamination from non-Activity dimensions in a number of cultures. In these cases, the recommended Activity scales might reasonably be supplemented or replaced, using procedures discussed below.

How does one proceed when working in a language community other than the 24 listed in Table 2.1? The cross-cultural project provided guidance with respect to this problem. Having established the validity of the Evaluation–Potency–Activity affective dimensions in more than 20 different language–community locales, it was deemed no longer necessary to go through the complete scale development process when expanding to new locales. The EPA structure may be assumed, leaving only the question of what scales should be used to measure each dimension in a new language. Researchers in the cross-cultural project addressed the issue in developing scales for German and Hebrew.

In the case of German, the following note preceded listing of Atlas data in the printout. "In this language/culture community the 'tool-making' stage was bypassed. Scales for the Atlas concept testing were inferred from the surrounding communities: Belgium (Flemish), the Netherlands (Dutch), and Sweden (Swedish)." In other words, the strategy for identifying adjective pairs in a new language was to extrapolate tool-making results from closely related language–culture communities.

In the case of Hebrew, a preliminary note in the printout said the following. "In this language/culture community the 'tool-making' stage was bypassed. Scales for the Atlas concept testing were translated from the four most frequent E, P, and A scales from the pan-cultural analysis." This simpler strategy assumed that adjective concepts that arose most often in the cross-cultural project can be extrapolated to any new language–culture community. The adjective pairs involved in such an approach are the following.

- *Evaluation:* good–bad, beautiful–ugly, pleasant–unpleasant, worthwhile/valuable–worthless
- *Potency:* big–little, strong–weak, heavy–light, high/tall–short
- *Activity:* fast–slow, young–old, active–passive, noisy–quiet

A third strategy might be to begin with the pancultural adjective pairs just listed and modify them judgmentally or add entirely new pairs that one thinks best reflect the underlying affective dimensions in the language of interest. However, scales developed creatively in this way must be checked to see if they truly measure the pancultural affective dimensions. This can be done by using the new scales to rate the following concepts, which are known to have extreme ratings panculturally on the three dimensions.

- *Negative evaluation:* murder, cancer, air pollution, crime, suicide, disease, poison, accident, war, hell, filth, sickness, cheating, earthquake, enemy, thief, hate, drunkenness, devil, atomic bomb. *Positive evaluation*: mother, friendship, peace, success, health, happiness, father, heaven, kindness, sense of sight, beauty, freedom, spring, love, food, eyes, library, nurse, flower, bath.
- *Negative potency:* baby, insect, child, rabbit, a second, chicken, bird, a point, one, flower, lipstick, miniskirt, cat, rose, grass, spit, beggar, cup, two, pocket radios. *Positive potency*: elephant, sun, mountains, airplane, atomic bomb, the universe, sea, dam, nuclear submarine, ship, army, a pyramid, train, lion, world, North America, Asia, university, one million, steel.
- *Negative activity:* rock, a pyramid, funeral, old people, stone, wall, deserts, prison, table, old age, mountains, floor, deaf, chair, silence, iron, solitude, blind, roof, wood. *Positive activity*: boy, youth, rabbit, adolescence, child, student, bird, play, automation, dog, miniskirt, champion, son, nurse, hero, horse, sense of sight, wife, space travel, I (myself).

A good approximation to pancultural factor scores for the newly created scales can then be obtained by correlating ratings on the new scales with the mean ratings for these concepts across 23 cross-cultural communities [given in Appendix F of the Osgood, May, and Miron book (1975)] or with mean ratings for 17 cross-cultural communities computable with the SPSS file provided at the ftp site associated with this book.

2.3 INTERNET DATA COLLECTION

The *Surveyor* data collection system for assessing sentiments via the Internet (Heise 2001a) can be revised to work in any indigenous language. Currently, renditions are available in Japanese, German, Spanish, simplified Chinese, and Arabic, as well as English. Its cross-linguistic capabilities, and other options, are covered in detail in the program documentation (Heise 2005).

The Evaluation scale that was used for data collections reported in this book was defined with the contrasts *good, nice* and *bad, awful*. Potency assessments were obtained on a scale with the adjectives *powerful, big* at one end and *powerless*, *little* at the other end. Activity was measured on a scale with adjectives *fast*, *noisy*, *lively* opposing *slow*, *quiet*, *lifeless*.

Respondents began a data-collection session by pointing their browsers to an Internet address that was provided to them. Upon clicking a button indicating that they understood and accepted the conditions of the study, they arrived at an Internet page that loaded the data-collection applet *Surveyor*. *Surveyor* first questioned the respondent about demographic and social matters. Survey items in closed-form format were presented below the instruction, "Click the mouse pointer on your answer, then click the 'Okay' button." The respondent clicked a radio button to answer each question, whereupon the Okay button worked to clear the screen and present the next display.

After background questions had been presented, the program presented a 12-step tutorial on how to record feelings about things on semantic differential scales such as the one displayed in Figure 2.1. In each step the screen was cleared, and the texts quoted below appeared on the screen, along with a clickable button (indicated in boldface below). The respondent had to interact with the tutorial to complete each step.

1. "In the rest of this project you rate your feelings about different concepts. A word or phrase says what the concept is. Click the **Next** button to see how a concept is presented."
2. The screen showed the stimulus "helping someone" and the following text. "You rate your feelings about the concept with a kind of ruler. The ruler has a slider that you can move from one extreme to the other. Click the **Next** button to see the ruler."
3. A horizontal bar and pointer appeared unembellished below the stimulus. "Practice rating now. Position the mouse pointer over the slider. Hold down the mouse button and drag the slider along the ruler. Release the mouse button above a mark, or between marks, to finish rating." The next screen appeared when the respondent released the mouse button.

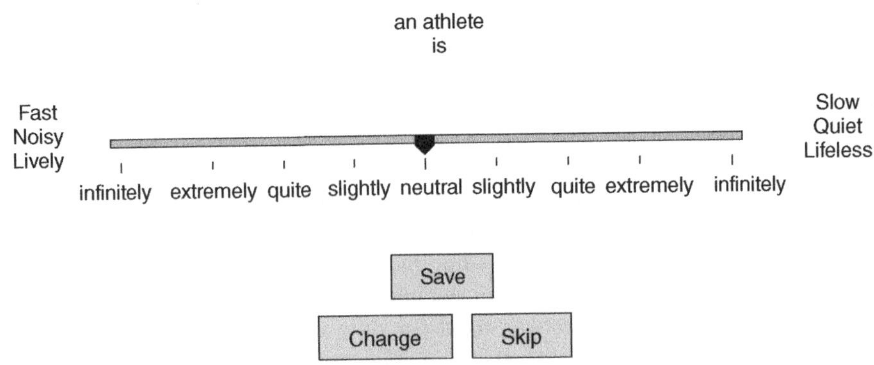

FIGURE 2.1 Screen display for the sentiment-measuring instrument.

4. Next, the scale was elaborated by displaying adverbial modifiers. "The ends of the ruler represent the most extreme conditions imaginable. The middle represents neither extreme; the middle is neutral. Click the **Next** button to see how ends of the ruler get defined."
5. The scale was elaborated further by displaying adjectives at either end. "Here the ends of the ruler represent good versus bad. (Sometimes the good and bad sides will be reversed on the scale.) Click **Next** to see another pair of extremes that you will use with every concept."
6. "These extremes let you rate whether the concept is powerful or powerless. Base your rating on your first impression, rather than on logical reasoning. Click **Next** to see a third pair of extremes used with every concept."
7. "These extremes concern your feelings about the liveliness of the concept. Rate "helping someone." Is the behavior typically lively or sluggish? Click the **Save** button when you have the slider where you want it."
8. "You must move the slider in order to do a rating. To rate something as neutral, move the slider away from the middle, then back. Practice this now by rating "a stranger" as neutral on this scale." The requirement mentioned was intended to minimize satisficing (Krosnick and Alwin 1987) among respondents with low motivation.
9. "To change a rating, click the **Change** button. Then a list will appear showing the concepts you have rated. Click the **Change** button now."
10. "When you click **Change**, a box lists the concepts that you have rated. You click on a concept you want to rate again. Or click **No change** to cancel. Now, though, click the **Next** button to learn one more thing."
11. The stimulus above the scale changed to "a polymath." "If a concept is unfamiliar to you, click the **Skip** button. Click **Skip** now (even if you know what "polymath" means). You will have to click another button, then wait a few seconds."
12. "Now you can begin rating concepts. Click the **Okay** button when you are ready."

After these instructions, respondents recorded their feelings about a set of concepts—the stimuli. *Surveyor* randomly selected a stimulus from the set of 100 stimuli to be rated.

The stimulus chosen was presented on the computer screen as shown in Figure 2.1. The set of adjectives defining the scale—for E, P, or A—was selected randomly. The orientation of adjectives on the scale was also randomized; for example, the Activity scale sometimes appeared with *fast–noisy–lively* appearing on the right side, and sometimes on the left. The scale consisted of a line drawn over 430 pixel positions, any pixel of which could be chosen as a rating. Ratings at the "neutral" center point of the scale were coded 0.0, and ratings elsewhere were coded proportionately to their distance from the center point, from –4.3 to +4.3. Adverbial quantifiers were used to characterize positions on the scale. In line with studies of adverbial quantification, the word "slightly" marked a 1-unit distance from the scale's center point; "quite" a 2-unit distance, "extremely" a 3-unit distance, and "infinitely" a 4.3-unit distance. Ratings on the *good*, *powerful*, and *lively* sides of scales were treated as positive; ratings on the opposite side, as negative.

After a few stimuli had been rated, a line appeared below the Save, Change, and Skip buttons, saying "Left to do:" followed by the number of stimuli the respondent still had to rate. When the respondent signaled completion of a rating by clicking the Save button, *Surveyor* coded and saved the pointer's graphic position, reset the pointer to the middle position, and presented the next EPA scale, chosen randomly, for rating the current stimulus. When that stimulus had been rated on all three dimensions, a new stimulus was chosen randomly. After the last stimulus, the Change option was invoked automatically so that the respondent could re-rate any stimulus.

Finally, the respondent filled in a form with a name and address so that a gratuity could be given. The name and address were sent (without data) in an email to the researcher, and then discarded. *Surveyor* bundled the respondent's data (without the name and address) and sent the packet over the Internet to an archive on a central computer server.

2.4 CHAPTER HIGHLIGHTS

- Sentiments are cultural concepts with affective associations. The affect varies along three bipolar dimensions: Evaluation—*goodness* versus *badness*; Potency—*weakness* versus *powerfulness*, and Activity—*quiescence* versus *activation*. Attitude measurements are assessments of sentiments on the Evaluation dimension only.
- The EPA dimensions are integral components of systems for classifying personalities, interpersonal behaviors, connotative word meanings, expressive displays of the body, and human emotions.
- Scales for measuring attitudes and sentiments can be constituted from items that characterize specific positions along a continuum or from items that are endorsed by different proportions of people, depending on their position on an issue. Respondents can also register their attitudes and sentiments with ratings on bipolar scales anchored at each end by adjective extremes.
- Empirical research revealed that the three dimensions of Evaluation, Potency, and Activity (EPA) underlie ratings on bipolar scales. A massive cross-cultural study, conducted with meticulous safeguards against ethnocentricity, demonstrated that the three dimensions of affect in bipolar ratings are cross-cultural universals.

- Bipolar rating scales provide a single measurement metric for all three EPA dimensions and for all kinds of stimuli. This standardization enabled the development of mathematical approaches to impression formation processes in humans.
- Graphic bipolar rating scales, implemented with computers, use distances on the scale to quantify ratings, thereby circumventing the problem of deriving a quantitative metric for response options.
- Anchoring the ends of bipolar scales with images is at present an impractical way of developing EPA scales in unexplored cultures. Instead, new scales with verbal anchors need to be composed and used to obtain indigenous ratings of select concepts. The scales whose ratings correlate with average cross-cultural ratings are reasonable instruments for measuring the EPA dimensions.
- An existing data collection system for assessing sentiments via the Internet can be revised to work in any indigenous language.

3 Sentiment Repositories

Dictionaries of word meanings and repositories of sentiments attached to words can function as galleries of subjective culture, allowing researchers to reconstruct significant components of sociocultural systems. For example, a 1623 dictionary defines *commotrix* as "A Maid that makes ready and unready her Mistris" (Landau 2001, p. 50), thereby linking three identities and referring obliquely to actions in the setting of a commode, which allows reconstruction of a piece of one social institution in Renaissance society (see MacKinnon and Heise 2010, Chapter 4). A listing of sentiments generated from ratings by 1970s' university students (Heise 1979, Appendix) indicates that the Evaluation–Potency–Activity (EPA) profile for mothers was 2.4, 1.4, 1.3 (very good, powerful, and active), and for babies was 1.6, –2.2, 2.5 (very good, very powerless, and very active), and by combining this information with similar information about behaviors and employing affect control theory's interaction simulation program (Heise 2007, Chapter 20) it can be inferred that normative actions of mothers toward babies for those respondents included the acts of flattering, stroking, admiring, amusing, and cuddling. In 1990s' Japan the sentiment for mother was 1.5, 1.2, –0.8, and for baby, 0.2, –2.6, 2.5 (Heise 1997), and similar analyses suggest that normative actions of Japanese mothers toward babies tended more toward teaching, protecting, charming, assisting, caressing, and soothing.

Thus, an important aspect of surveying cultures is creating repositories of denotative and affective meanings that can be used in characterizing the sociocultural system and comparing it with other sociocultural systems. In the first part of this chapter I review the history of repositories of affective meanings or sentiments, a task that is relatively manageable since the technology for producing sentiment repositories is only about half a century old (see Chapter 2). In the last part of this chapter I describe a recent project that compiled sentiments for 2,000 concepts. The project is particularly important in this book because it produced the dataset used for analyses that are reported in Chapters 6, 7, and 8.

Surveying Cultures: Discovering Shared Conceptions and Sentiments, By David R. Heise

3.1 EARLY ARCHIVES

Three sentiment repositories were published before 1970. Jenkins, Russell, and Suci (1958) reported semantic differential mean ratings of 360 words on 20 scales. About a fifth of the words chosen related to psychological studies of free association and meaningfulness; other words were chosen in an effort to find words with extreme semantic differential profiles, or to replicate measurements made previously by Charles Osgood and colleagues at the University of Illinois (Osgood, Suci, and Tannenbaum 1957); other words were chosen because they "appeared interesting for a variety of reasons."

Selection of scales reflected the uncertainty at that time about the number of reliable factors influencing semantic differential ratings: "The scales were selected to sample six factors: 'Evaluation'—eight scales sampling four subcategories of this broad, pervasive factor; 'Potency'—three scales; 'Activity'—three scales; Tautness'—two scales; 'Novelty'—two scales; and 'Receptivity'—two scales" (Jenkins, Russell, and Suci 1958, p. 690). Respondents were 270 women and 270 men volunteering from an introductory psychology course. Eighteen questionnaire booklets were prepared, each containing 20 words and distributed to 30 respondents.

The Jenkins, Russell, and Suci study included a small reliability study of 20 concepts drawn randomly from the list of 360 words and re-rated four weeks after the main data collection. They found high duplicability across all scales as applied to the 20 concepts: "the test–retest reliability of mean scale values (20 scales × 20 concepts, or a scatter plot of 400 entries) yields a Pearson r of 0.97" (1958, p. 693).

Jenkins, Russell, and Suci did not compute factor scores but instead reported the mean rating of each concept on each scale. For example, the first line of their listing is "ABORTION 160 370 573 332 273 573 277 217 637 493 643 267 343 382 567 283 343 553 647 430" (1958, p. 695), which represents the mean ratings (times 100) of abortion on the scales *cruel–kind*, *curved–straight*, *masculine–feminine*, *untimely–timely*, *active–passive*, *savory–tasteless*, *unsuccessful–successful*, *hard–soft*, *wise–foolish*, *new–old*, *good–bad*, *weak–strong*, *important–unimportant*, *angular–rounded*, *calm–excitable*, *false–true*, *colorless–colorful*, *usual–unusual*, *beautiful–ugly*, and *slow–fast*, respectively. The seven-step scales were coded 1.0 to 7.0, so 4.0 has to be subtracted (after dividing by 100) in order to convert to the usual minus–plus values for semantic differential scales. For example, abortion's most extreme ratings were cruel (–2.40), foolish (2.37), bad (2.43), and ugly (2.47).

The second published repository of sentiments was my "Semantic Differential Profiles for 1,000 Most Frequent English Words" (1965). The focus on frequent words (apart from function words such as articles and prepositions) came from my goal of using the EPA profiles as a basis for content analyses (Heise 1966). Selection of words was based on frequencies of semantic units rather than on raw word frequencies, and about 50 additional words were included because of their special relevance in the content analysis project. Each word was rated on eight scales, representing four dimensions:

Evaluation (*good–bad, pleasant–unpleasant*), Potency (*strong–weak, tough–tender*), Activity (*active–passive, lively–still*), and Stability (*rational–emotional, tamed–untamed*).

The study took an unusual step in clarifying each semantic unit by including a brief illustrative sentence.

> Pilot work had indicated that definition of word concepts could not be achieved through the use of synonyms since the mere presence of other content words contaminates the affective connotation of a stimulus word. Pilot work also had indicated, however, that presence of function words has relatively little effect on the affective connotation of a stimulus. Thus it was feasible to define each word concept by giving an example of its use in a sentence composed otherwise of function words only. A 67-word vocabulary of function words was used in constructing defining sentences; these words alone sufficed to define 90.9 percent of the 1,000 semantic units on the list (for 91 entries, use of a nonfunction word was required to make the concept's meaning clear). . . . Verbs were defined by sentences in which the verb was used in the simple past; nouns were de-fined by sentences in which the noun was used in the singular (except in a few cases where this seemed awkward and opposed to common usage). (Heise 1965, p. 4)

For example, *mother* was defined by the sentence "It is his mother"; and the verb form, had it been included, would have been defined with "She mothered him." Some special analyses in the study determined that "the SD ratings presented in the dictionary would be substantially the same even if: (a) verbs had been defined using some tense other than the simple past; (b) nouns had been defined in the plural form" (Heise 1965, p. 10).

Respondents were Navy enlistees (average age, 18.9) training at the Hospital Corps School, Great Lakes Naval Training Center in Illinois. I administered 11 one-hour sessions with groups of these respondents in the spring of 1963, as a 26-year-old civilian. Each respondent rated 50 different words, and each word was rated by 16 different respondents. Factor analyses indicated that the supposed Stability scales actually loaded on Evaluation (*tamed–untamed*) and Potency (*emotional–rational*), so the Stability dimension was jettisoned, and those scales used instead in factor score equations to compute mean EPA profiles.

A third compendium of 550 sentiments collected from high school boys in Urbana, Illinois, was published as an appendix to James Snider and Charles Osgood's sourcebook on the semantic differential (1969). Snider and Osgood (1969, p. 625, footnote) indicated that the data were collected as part of the worldwide cross-cultural study described in Chapter 2, presumably representing the White American English repository of EPA profiles collected at the final stage of the cross-cultural project in each venue (see Section 3.2). However, the number of concepts was fewer than the 617 listed in the final compendium for white American English, and Snider and Osgood's list contained a number of special-interest concepts that were not included in the compendiums for other societies (brown race, concept, dirt, evolution, the frug, high school, magic, menstruation, the NAACP, my parents, peace corps, pornography, sit-ins, thin, Viet Nam). Overall, the Snider and Osgood compendium contains a broad range of concepts in a variety of categories, such as time, kinship, society, and philosophy.

I collated materials from the foregoing three studies into a single list of 1,551 sentiments (Heise 1978, Appendix B). EPA profiles were converted to a uniform metric geared to the Snider and Osgood measurements, and redundant profiles for concepts that had been rated in multiple studies were averaged. The results were listed with concepts separated by grammatical type. This procedure resulted in a list of 168 possible actors and 48 interpersonal verbs that were used in early impression-formation research (see Chapter 4) and in the development of affect control theory (Heise 1977).

3.2 CROSS-CULTURAL ATLAS

The cross-cultural project described in Chapter 2 validated the universality of the Evaluation–Potency–Activity structure of affect. Additionally, the project produced a large corpus of information regarding each of the communities that had been studied. Charles Osgood and his colleagues referred to the collected results from the project as a cross-cultural *Atlas*. Materials in the *Atlas* included the qualifiers elicited from 100 substantives, along with indices assessing the ubiquity of each qualifier, various factor analyses of ratings made on scales constructed from the qualifiers, a set of indigenous scales for measuring EPA in each culture, and average EPA ratings using these scales to rate 620 concepts that had been selected to be familiar in all of the cultures studied. As Osgood, May, and Miron (1975, p. 241) said: "This is not in any sense a 'world atlas,' but nevertheless it has proven to be a complex and time-consuming endeavor."

The mean EPA profiles for 620 concepts in more than 20 language–culture communities rated by 40 males in their middle teens from schools of the relevant community (Osgood, May, and Miron 1975, p. 245) constituted a major accumulation of cross-cultural sentiment measurements. Osgood, May, and Miron (1975, p. 254) summarized the major considerations in choosing the 620 concepts as follows.

> In selecting concepts for the Atlas, we wished to get as much diversity as possible and yet have adequate representation of as many categories as we could—all within a manageable total number. We wanted concepts that would be intrinsically interesting, that would be potentially differential among cultures, that might tap human universals in symbolism, but that would also sample those everyday aspects of human life—kinship, foods, animals, technologies—which ethnographies usually record. The resulting collection of items of subjective culture runs from A to Z, however, as any good Atlas should—from ACCEPTING THINGS AS THEY ARE, ACCIDENT, and ADOLESCENCE through MARRIAGE, MASCULINITY, and MASTER to YESTERDAY, YOUTH, and ZERO.

The concepts tapped 12 general categories (Osgood, May, and Miron 1975, Appendix G): time, kinship, abstract symbolisms, concrete symbolisms, environmentals, carnalities, human activity, interpersonal relations, society, communications, philosophy, and things and stuff. Most categories were further divided into subcategories which contained multiple concepts. For example, abstract symbolisms was subdivided into emotions, numbers, colors, geometricals, and days; and days contained the days of the week plus the concepts of "day" and "week."

Osgood, May, and Miron (1975, Preface, note 5) anticipated that a future volume, *The Affective Dimensions of Subjective Culture*, would report on *Atlas* data. The volume would have contained all of the acquired materials, judging from another of their comments (Osgood, May, and Miron 1975, p. 193, note 1): "Although *Atlas* tables are available for the 23 communities which have reached this stage at the time of this writing, publication of the *Atlases* will be delayed until all communities now underway (a total of about 26) have reached this stage." Unfortunately, the *Atlas* was never published—evidently because the project collapsed after Osgood became ill.

I acquired some *Atlas* tables in the form of computer printouts that I purchased at the University of Illinois Bookstore in about 1978 (Heise 2009). The 17 language–culture communities in the print–outs were: Arabic (Beirut), Bengali (Calcutta, India), Dutch (Amsterdam and Haarlem), English (Illinois whites and blacks), Farsi (Teheran), German (Münster), Hebrew (Israel), Hindi (Delhi, India), Malay (Kelantan state), Portuguese (Portugal), Serbo-Croat (Belgrade), Spanish (Mexico City, Yucatan, Costa Rica), Thai (Bangkok), and Turkish (Istanbul). Osgood, May, and Miron (1975, Table 5.2) list 29 language–culture communities that were considered a possible part of their cross-cultural project at the time of their 1975 book, and 14 of those communities were not in the computer printouts that I purchased: Cantonese (Hong Kong), Dari (Kabul, Afghanistan), Finnish (Helsinki), Flemish (Brussels), French (Paris, Strasbourg), Greek (Athens), Italian (Padua), Japanese (Tokyo), Kannada (Mysore City and Bangalore, India), Magyar (Budapest), Pashtu (Kabul and Kandahar, Afghanistan), Polish (Warsaw), Swedish (Uppsala), and Tzeltal (Chiapas, Mexico).

The Preface tells how image files of the printouts that I purchased in the 1970s can be downloaded. To my knowledge, all *Atlas* materials for the 14 communities not in the computer printouts are lost, except for a few selected tables presented in Osgood, May, and Miron's (1975) book.

3.3 ARCHIVES RELATED TO SOCIAL INTERACTION

From the 1970s forward, repositories of cultural sentiments were assembled almost exclusively by sociologists working on affect control theory (ACT; Heise 2007) for use in impression-formation studies (see Chapter 4 of this book) and to support simulations of social interactions based on affect control theory (Schneider and Heise 1995). Interest in this sociological theory of social interaction put a distinctive stamp on the composition of these archives. Each repository contained Evaluation–Potency–Activity profiles for a large number of social identities (i.e., names for different kinds of people) and for a large number of interpersonal behaviors. After the early years, each corpus also contained EPA profiles for standard emotion terms, and many also contained words designating personal attributes of individuals. Many of the later datasets also contained EPA profiles for social settings.

As of 2009, repositories of cultural sentiments related to affect control theory have been collected in six cultures: United States, Canada, Ireland, Germany, Japan, and China. These repositories can be exported from the *Interact* social interaction simulation program (Heise 1997).[1]

3.3.1 United States

Three datasets were collected from American respondents in the years 1975 to 1998. A fourth dataset is described in Section 3.4.

North Carolina, 1975. I obtained EPA ratings of 650 social identities and 600 social behaviors from University of North Carolina undergraduates in 1975, using paper questionnaires. Scales were defined by the following adjectives: Evaluation—*bad, unhelpful, awful, sour* versus *good, helpful, nice, sweet*; Potency—*weak, powerless, little, shallow* versus *strong, powerful, big, deep*; Activity—*slow, dead, quiet, stiff, old* versus *hurried, alive, noisy, fiery, young*. A metric for the nine rating positions was derived by successive-intervals scaling (Heise 1978, Chapter 4).

The selection of concepts in this study is of interest since the study's lexicon became a basis for selecting concepts in later projects as well. I described the procedure as follows (Heise 1978, p. 63):

> Lists of identities and interpersonal acts were created by examining every entry in the *Doubleday Dictionary* (Landau 1975), a relatively short dictionary (906 pages) explicitly developed to cover a broad range of general vocabulary including informal usages, slang, obscenities, and vulgarisms. Slang and argot dictionaries were also examined but they were found to contain fairly few useful words relative to their esoteric and out-of-date entries.
>
> The verb list consists of transitive interpersonal verbs designating acts that people can do to other persons. About 1,200 such verbs were identified (excluding multiple-word verbs, like "turn to"). However, by selecting just the words that might define how laymen view social interactions; by excluding odd, esoteric words or references to rare acts; and by deleting entries whose sense is unclear out-of-context (for example, "to beat someone"), the list was reduced to 600 entries.

[1] A translation of the *Surveyor* interface to Arabic was achieved in 2009 by Muhammad Abdul-Mageed; ratings are to be obtained in Middle Eastern locales in 2010 and 2011.

> Thousands of entries in the *Doubleday Dictionary* refer to social identities, but names of noncontemporary roles and person-labels created from verbs plus "er" (for example, "corrupter") were ignored. Attention was focused on particular social domains, and only words referring to statuses and roles in these domains were included. Thus the final set of 650 nouns is rich in social identities associated with *courtrooms, hospitals, stores, families, classrooms, entertainments, football, peer groups, sexuality*, and the *underworld.*

Respondents were 311 undergraduates, 61 percent of whom were females. Ten different questionnaires were distributed, and the number completing each questionnaire varied from 28 to 34. EPA profiles were averaged just for females, with male increments provided when significant. This repository appears in printed form as an appendix in each of my 1970s books relating to affect control theory (Heise 1978, 1979).

North Carolina, 1978. In a National Institute of Mental Health–funded project, Lynn Smith-Lovin and other graduate students under my direction used paper questionnaires to acquire ratings of 721 identities, 600 behaviors, 440 modifiers, and 345 settings.[2] Ratings were obtained from 1,225 North Carolina undergraduates. The number of male or female raters generally was about 25 for each word.

Scales were defined by the following sets of adjectives: Evaluation—*bad, awful* versus *good, nice*; Potency—*little, powerless* versus *big, powerful*; Activity—*slow, old, quiet* versus *fast, young, noisy*. Scale values for the nine rating positions were derived by successive-intervals scaling (Smith-Lovin 1987b; Smith-Lovin and Heise 1988, pp. 42–45).

Identities and behaviors were largely the same as in the 1975 North Carolina study, with the identities list supplemented by names of occupations. Modifiers were selected on the basis of prior research on trait attribution and on moods, plus some labels for social characteristics such as rich and old (for details, see Averett 1981, pp. 34–35). This being the first study oriented explicitly toward assessing sentiments associated with settings, the *Doubleday Dictionary* was searched for places large enough for at least two people to interact, including pathways and vehicles, resulting in 1,274 place names, 112 paths or boundaries, and 277 vehicles. These were combined with the names of 303 recurrent situations or events (e.g., Christmas), 203 settings defined by a group of people (e.g., a mob), and 96 multiword settings that were not in the dictionary (e.g., a doctor's office), yielding a list of 2,265 settings from which 345 were selected for the compendium of setting sentiments (for more details, see Smith-Lovin 1987a, pp. 79–80).

[2]Ratings for a few emotion words in this dataset were obtained from Indiana University undergraduates in 1985, to accord with a systematic definition of the emotion lexicon (Ortony and Clore 1981; Ortony, Clore, and Foss 1987) published after the data collection in North Carolina.

Texas, 1998. Andreas Schneider collected ratings of 443 identities, 278 behaviors, 65 modifiers, and one setting at Texas Tech University with computerized graphic rating scales (program *Attitude*—see Chapter 2). The 482 respondents were nearly equally male and female, and approximately 30 of each gender rated each concept. Concepts were largely the same as in the 1989 Germany study (see below).

3.3.2 Canada

A number of data collection projects in Canada were conducted by Neil MacKinnon with funding from the Social Science and Humanities Research Council of Canada.

Ontario, 1980–1986. Data on 843 identities and 593 behaviors were obtained from 5,534 Guelph, Ontario, undergraduates with paper questionnaires in 1980–1983, and 495 modifiers rated by 1,260 Guelph undergraduates were added in 1985–1986. The number of males and the number of females rating each concept typically were 25 to 30.

Scales were the same as in the North Carolina, 1978 study. Metrics defining the values of choice points on the nine-position scales were derived by successive-intervals scaling of Canadian data. Concepts largely matched those considered in the 1978 North Carolina study; additional concepts were included that related particularly to Canadian society.

Ontario, 2001–2003. Data on 993 identities, 601 behaviors, 500 modifiers, and 200 settings were gathered from Guelph, Ontario, undergraduates in 2001–2002 with the computerized graphic rating scales of program *Attitude*. Data on settings were gathered at Guelph in 2003 with the *Surveyor* data collection system. (See Chapter 2 for details on the *Attitude* program and the *Surveyor* system.) This study was intended as a replication of the 1980s study and therefore assessed the same concepts, plus concepts relating to various ongoing studies.

3.3.3 Ireland

Belfast, 1977. Dennis Willigan, with guidance from myself and funding from the Jesuit Council for Theological Reflection, obtained ratings of 528 identities and 498 behaviors from 319 Belfast teenagers in Catholic high schools in 1977, using paper questionnaires. Ratings of modifiers and settings were not obtained in the Belfast study. Up to 18 females and 14 males rated each concept.

Scales were the same as were used in the North Carolina, 1975 study. Concepts were chosen in part to replicate the North Carolina, 1975 study. Beyond that, selections of concepts were oriented toward reflecting Irish society and the conflict with Britain.

3.3.4 Germany

Two repositories of sentiments have been compiled in German.

Mannheim, 1989. Andreas Schneider, with my guidance, collected ratings of 442 identities, 295 behaviors, and 67 modifiers, using his translation of the *Attitude* program into German. Subjects were 520 Mannheim students, matched to American undergraduate populations by proportional inclusion of youths in grades 12 and 13 at two German Studenten des Grundstudiums and Gymnasiasten, along with subjects from Mannheim University, which attracts students mainly from the Rhein–Neckar region. Approximately 30 males and 30 females rated each concept.

Scales were defined by the following sets of qualifiers: Evaluation—*angenehm, gut, freundlich, schön* versus *unangenehm, schlecht, unfreundlich, hässlich*; Potency—*klein, leicht, zart, schwach* versus *gross, schwer, kraftvoll, stark*; Activity—*bewegt, lebhaft, geräuschvoll, schnell* versus *ruhig, gemessen, still, langsam*.

Concepts were selected from the 1978 North Carolina study, with particular emphasis on back-translatability between German and English.

Internet, 2007. Tobias Schröder assembled EPA profiles for 1,100 words: 376 social identities, 393 social behaviors, and 331 adjectives denoting personality traits and emotional states.

Data were collected via the Internet with the *Surveyor* system. Respondents were 734 males and 1,171 females from all over Germany who chose to participate in a "study of language and emotion," in response to an extensive publicity campaign conducted via mailing lists in different universities, Web logs, newspaper reports, and radio interviews. Most of the participants (N = 1,029) were between 20 and 29 years of age, but the sample covered all ages, including 129 respondents younger than 20 and 92 older than 60 years.

Adjectives for defining the Evaluation, Potency, and Activity bipolar scales were the same as in the Schneider study. An effort was made to select the most important German words in describing social interaction, and also to replicate many of the concepts in the Schneider compendium.

3.3.5 Japan

Japan, 1989–2002. A group of sociologists—Herman Smith, Shuuichirou Ike, Takanori Matsuno, and Michio Umino—compiled ratings of 403 identities and 307 behaviors, and a few settings from 323 Tohoku University students in 1989, using the *Attitude* program (translated into Japanese by Shuuichirou Ike). In 1995 and 1996, 120 women students at Kyoritsu Women's, Japan Women's, and Teikyo universities and 120 men students at Teikyo and Rikkyo universities rated an additional 300 settings, 300 modifiers (mainly traits), 200 business identities, and 75 behaviors. With the aid of Yoichi Murase and Nozomu Matsubara, students at Rikkyo University and Tokyo University rated 102 emotions, 70 behaviors, and 55 identities in 2002 using the *Surveyor* system. Total numbers of entries in the Japanese sentiment repository are: 713 identities, 455 behaviors, 426 modifiers, and 300 settings. The number of male or female raters generally was about 30 for each concept.

> Our lexicons have an intentionally wide overlap with American affect control theory dictionaries. Eighty-two percent of the verbs and 47 to 53 percent of the identities back-translate blindly into English entries. The identities and actions in our dictionaries accord with broad groupings (legal, medical, family, school, crime, and intimate) for breadth of coverage. The Japanese identities and actions cover the entire spectrum of goodness, powerfulness, and liveliness. All the choices of stimuli are well within the normal language comprehension of adult Japanese subjects. (Smith, Matsuno, and Umino 1994, p. 128)

The Japanese graphic rating scales were anchored at each end with qualifiers as follows: Evaluation—*yoi, rippa na* (good, nice) versus *warui, hidoi* (bad, awful); Potency—*tsuyoi, chikara no aru, ookii* (strong, powerful, big) versus *yowai, chikara no nai, chiisai* (weak, powerless, little); Activity—*hayai, wakawakashii, sawagashii* (fast, young, noisy) versus *osoi, ochitsuita, shizuka na* (slow, old, quiet).

3.3.6 China

China, 2001. With funding from the National Science Foundation, Herman Smith and Yi Cai obtained ratings of 449 identities, 300 behaviors, 98 emotions, 150 traits, and 149 settings from about 380 undergraduate students at Fudan University in Shanghai, People's Republic of China. Yi Cai translated the interface of the *Attitude* program into a Chinese version called *Taidu*, and that was the basis of data collection.

Anchors for the Evaluation scale were: *good* and *nice* (*haode, lingrenyukuaide*) versus *bad* and *awful* (*Huaide, zaogaode*); for Potency the anchors were *big* and *powerful* (*dade, youli de*) versus *little* and *powerless* (*xiaode, wulide*); and for Activity the anchors were *lively, fast*, and *young* (*chongmanhuolide, kuaide, zhaoqipengbode*) versus *quiet, slow*, and *old* (*anjingde, chihuande, muqichenchende*).

Later studies collected data with the *Surveyor* program. In this case anchors for the Evaluation scale were: *good, lovely* (*hao, ke ai*) versus *bad, detestable* (*huai, ke zeng*); for Potency: *strong, forceful* (*jian qiang, You li*) versus *soft* and *weak, powerless* and *weak* (*ruan ruo, wu li*); for Activity: *excited* and *spirited, lively* and *excited* (*xin qing ji ang, huo po xing fen*) versus *calm* and *sober, depressed* and *gloomy* (*xin qing chen wen, ya yi*).

Work on Chinese culture continued in collaboration with sociologists Luo Jar-Der and Wang Jin, until Smith's death in 2007. Wang Jin continues work on the project.

3.4 U.S. 2002–2004 PROJECT

At the turn of the century two large repositories of sentiments were available from the United States: one compiled in North Carolina in 1978 by Lynn Smith-Lovin and myself which covered 2,106 concepts, and one compiled in Texas in 1998 by Andreas Schneider which covered 787 concepts. Although the Texas study provided relatively contemporary measurements, it had sparse numbers of behaviors, person modifiers, and social settings. Therefore in 2001, I, in cooperation with Clare Francis, compiled an update of the 1978 study with a broad range of identities, behaviors, modifiers, and settings related to social interaction. This twenty-first-century update is the source of empirical data in this book.

Ratings of the concepts on semantic differentials (Heise 1969c; Osgood, Suci, and Tannenbaum 1957) were collected over the Internet in three endeavors that recruited respondents from different pools: (1) Arts and Sciences students at Indiana University—these respondents provided one round of ratings; (2) Business School students in an introductory management course at the same university—these respondents provided two rounds of ratings, including the test–retest ratings analyzed in Chapter 7; and (3) Arts and Sciences students at the University of Connecticut—one round of ratings. Ratings of the concepts were obtained from all three samples, along with some sociodemographic information about the respondents themselves. The Business School students additionally rated some course-related concepts in their first wave of participation in early September 2003—hereafter called *time 1*. Additional semantic differential ratings including re-ratings of eight identities and eight behaviors, plus other course-related measurements, were obtained from the Business School students in a second wave of data collection from the middle of October to the end of November 2003—hereafter called *time 2*.

3.4.1 Stimuli

The focus of data collection consisted of 500 social identities, 500 social behaviors, and 200 social settings drawn from the social institutions of family, religion, medicine, education, law, politics, and business; concepts related to informal social relations and sexuality were also included. Three hundred personal modifiers included about 100 emotion terms, plus some additional cognitive–affective descriptors and other attribute terms unrelated to affectivity. Procedures for selecting these stimuli and lists of the stimuli are provided in an online article (Heise 2001a).

Individual respondents could devote no more than an hour to the rating task. Since rating all 1,500 stimuli would take seven to 10 hours, the list was divided into 15 subsets of 100 stimuli. Identities, behaviors, settings, and modifiers were distributed randomly and about equally among the subsets. Stimuli subsets were assigned randomly to respondents, and 64 or more respondents received each subset.

TABLE 3.1 Test–Retest Stimuli

Desired Evaluation–Potency–Activity Profile	Identities	Behaviors
Good, powerful, lively	a teammate	to sell something to someone
Good, powerful, quiet	a scientist	to request something from someone
Good, powerless, lively	an intern	to wait on someone
Good, powerless, quiet	a retiree	to whisper to someone
Bad, powerful, lively	a racketeer	to boss around someone
Bad, powerful, quiet	an executioner	to bill someone
Bad, powerless, lively	a salesman	to bicker with someone
Bad, powerless, quiet	a chain smoker	to murmur to someone

Identities and behaviors to be re-rated at time 2 were selected systematically to represent all patterns of sentiments and to favor concepts relevant to Business School students. The patterns and the concepts approximately representing each pattern are given in Table 3.1.

3.4.2 Auxiliary Questions

The following questions preceded the rating task in the questionnaire administered to all respondents. Percentages of respondents are given after answer categories, and *N*s are given after the question. Foreigners were excluded from analyses in this book, so *N*s do not include respondents who lived outside the United States before entering college.

Are you: (1113)
- Female (52.8%)
- Male (47.2%)

What is your dominant background? (1113)
- White (86.0%)
- White Hispanic or Latino (0.3%)
- Black or African-American (2.2%)
- Black Hispanic or Latino (0.4%)
- American Indian or Native American (0.2%)
- Native American Hispanic or Latino (0.3%)
- Asian or Asian-American (6.6%)
- Other (2.2%)

Answers to this item were coded as Asian or Asian-American, black or African-American, Hispanic or Latino (white, black, or Native American), Other (including American Indian or Native American), and white.

Where in the United States did you mainly live prior to entering college? (1113)

New England = ME VT NH MA CT RI (10.1%)
Middle Atlantic = NY NJ PA (8.4%)
East North Central = WI IL IN MI OH (67.7%)
West North Central = MN IA MO ND SD NE KS (5.1%)
South Atlantic = DE MD WV VA NC SC GA FL DC (1.7%)
East South Central = KY TN AL MS (2.4%)
West South Central = AR OK LA TX (1.6%)
Mountain = MT ID WY NV UT CO AZ NM (0.5%)
Pacific = WA OR CA AK HI (2.5%)
Not in United States (__)

Answers to this item were coded Northeast (New England and Middle Atlantic), Central (East North Central and West North Central), South (South Atlantic, East South Central, or West South Central, West (Mountain and Pacific). Respondents who answered "Not in United States" are not included in analyses in this book.

What is your marital status? (1113)

Never married (97.6%)
Married now (2.0%)
Widowed (0.1%)
Divorced (0%)
Separated (0.1%)

Answers to this question were coded Never married versus Married now or previously.

Additional background information was obtained in the time 2 questionnaire administered to Business School respondents, as follows.

Your age? (713)

20 or less (56.0%)
21 to 25 (43.3%)
26 to 30 (0.4%)
31 to 40 (0.3%)
over 40 (0%)

Answers to this question were coded 20 or less versus 21 or more.

Your part-time work experience? (713)

None (3.1%)
Less than 6 months (10.6%)

6 to 12 months (9.3%)
13 to 18 months (6.4%)
19 to 24 months (5.2%)
More than 24 months (29.6%)

Your full-time work experience? (713)
None (34.5%)
Less than 6 months (26.9%)
6 to 12 months (18.9%)
13 to 18 months (7.3%)
19 to 24 months (3.8%)
More than 24 months (8.6%)

Your current estimated GPA? (713)
Less than 2.0 (0%)
2.0 to 2.5 (3.6%)
2.6 to 3.0 (28.8%)
3.1 to 3.5 (52.9%)
3.6 or higher (16.7%)

The first two answer categories of this question were combined for analyses.

For the next four questions, the original answer categories were Strongly agree, Agree, Neither agree nor disagree, Disagree, and Strongly disagree. Agree and Strongly agree categories were combined; and the Disagree and Strongly disagree categories were combined; the percentages below are for the combined categories.

I find the process of learning new material fun. (642)
Agree (82.7%)
Neither agree nor disagree (14.5%)
Disagree (2.8%)

I browse in the library even when not working on a specific assignment.
Agree (24.9%)
Neither agree nor disagree (17.1%)
Disagree (57.9%)

I will withdraw from an interesting class rather than risk getting a poor grade.
Agree (25.7%)
Neither agree nor disagree (20.2%)
Disagree (54.0%)

I cut classes when confident that lecture material will not be on an exam.
  Agree (27.7%)
  Neither agree nor disagree (24.6%)
  Disagree (47.7%)

How many groups did you participate in this semester—social; athletic; religious; political; etc.? (642)
  None (6.1%)
  1 (9.8%)
  2 (21.2%)
  3 or 4 (41.9%)
  5 to 8 (19.3%)
  9 or more (1.7%)

The last two answer categories of this question were combined for analyses.

3.4.3 Performance Variables

Besides recording a respondent's ratings and answers to questions, the Java applet recorded the number of stimuli that the respondent skipped and the number of minutes that the respondent spent with the sentiment-rating program, including time spent on background questions and the tutorial.

Prior to eliminating foreign respondents and respondents who skipped more than 30 stimuli, the Business School respondents numbered 828. Within this group, the number of skipped stimuli ranged from none to 100, most respondents skipped between 0 and 14 (the first and ninth deciles), and the median number skipped was three. The percentage of respondents skipping more than 30 stimuli was 1.6%.

In the same group, the number of minutes that a respondent remained in the sentiment-rating program ranged from eight to 3,084 (i.e., one respondent kept the program active on his computer for several days). Most respondents spent 18 to 48 minutes with the program (the first and ninth deciles), and the median was 27 minutes. Five percent of the respondents kept the program active for more than an hour.

3.4.4 Raters

Arts and Sciences students at Indiana University in 2002–2003 were recruited through a classified ad in the campus newspaper and through a flyer handed out in sociology classes. Usable data were obtained from 270 students.

The classified ad read as follows: "IU students, earn $5.00 by rating concepts over the Internet. See information page at www.indiana.edu/~socpsy/". The ad, costing $35 a week, recruited about 15 respondents per week.

Examining connection statistics, I discovered that I was losing almost 40% of the potential respondents at the point where I asked them to provide their university user name and password before starting the task. I eliminated this requirement, and responses went up to about 25 per week. However, after about two months of advertising, the response to the ad dried up almost completely.

I then created a flyer and asked instructors teaching the introductory course in sociology to distribute the flyer to their students. This was late in the semester and attendance seemed to be poor (about 60%), so I got somewhat less than 500 flyers into students' hands. The response rate was about 20%. The flyer was titled SOCIAL SCIENCE WANTS YOU and showed a character pointing at the reader—an appealing lady (my wife) rather than Uncle Sam. The text went on as follows:

> Help answer questions like these:
>
> - What is the expected behavior of mothers, fathers, daughters, and sons?
> - Do males and females expect different behavior of employers? Of employees?
> - What emotions result from being cheated, robbed, assaulted, molested?
>
> IT'S EASY TO HELP!
>
> - Get on the Internet and go to the study information page at www.indiana.edu/~socpsy/.
> - Click the button at the bottom of the page if you accept the agreement. You provide your data over the Internet.
> - Answer a few questions about your background, then rate 100 concepts like mother, employee, robbing, etc. The ratings take 20 to 60 minutes.
> - Finally, enter your name and address to receive a $5.00 Kroger gift certificate that you can use in the grocery, beverage store, pharmacy, cosmetics shop, etc.

A large group of Business School students at Indiana University was recruited from an introductory management course given in the fall of 2003. Participation was required as part of a methodology practicum in the course. These respondents completed two surveys. The first survey, early in the semester, presented demographic questions and 100 stimuli for rating on Evaluation, Potency, and Activity. The second survey, late in the semester, presented course-evaluation questions, questions about academic and extracurricular activities, and 16 stimuli for rating on Evaluation, Potency, and Activity. Personal information was collected from respondents in both surveys so that their ratings of stimuli at different times could be linked. Some of the 828 recruits were dropped from analyses because they were nonindigenous to the United States or they did not complete the time 2 survey. Additional respondents who skipped more than 30 stimuli in the two administrations combined were dropped, giving a sample size of 722 for test–retest analyses across respondents. Respondents who skipped more than 20 stimuli combined were dropped for analyses across concepts, yielding 713 respondents.

Northeastern Arts and Sciences students were recruited from introductory sociology courses at the University of Connecticut during the academic year 2003–2004. Students were given extra credit points for their participation in the study. Usable data were obtained from 130 respondents.

3.5 CHAPTER HIGHLIGHTS

- Researchers began compiling repositories of sentiments in the 1950s. All of the early repositories were obtained with American respondents, and each repository addressed special interests of the collectors.
- The cross-cultural project described in Chapter 2 obtained mean EPA profiles for 620 concepts from 40 teenaged males in 31 language–culture communities. The data are still available for 17 language–culture communities, but data for 14 communities have been lost.
- Affect control theory researchers have assembled repositories of EPA profiles for social identities, interpersonal behaviors, emotion terms, personal attributes, and social settings. Four such repositories have been collected from American respondents, two from Canadians, two from Germans, one from Irish, one from Japanese, and one from Chinese.
- Empirical analyses in this book employ ratings of 1,500 stimuli, partitioned into 15 sets, each set distributed to 64 or more respondents. Test–retest analyses of reliability employ ratings of 16 of the 1,500 stimuli, chosen to represent diverse sentiments.
- Background questions presented to all respondents determined their gender, ethnicity, geographic origin, and marital status. A large subset of respondents answered additional questions regarding their age, grade-point average, academic attitudes and behaviors, group affiliations, and work experience.
- The sentiment-measuring instrument collected respondent ratings of stimuli on bipolar graphic rating scales, measuring three dimensions of affective meaning with high levels of discrimination. Respondents were tutored in use of the instrument. Randomization was used extensively in presenting stimuli and scales to respondents. Respondents were allowed to skip stimuli and to change ratings already given.
- The median amount of time that respondents required to answer background questions, take the tutorial, and rate 100 stimuli on three dimensions was 27 minutes.
- The college student respondents were recruited in two ways: through promotions that promised volunteers a small remuneration; and through course requirements.

4 Surveys with Vignettes

Respondents cannot answer every question that we wish to put to them about their culture, because some of their cultural processing is unconscious, even while being normative and essentially the same from one person to another. People might, for example, perform poorly in ranking the importance of various relatives and be even worse at identifying factors that make one relative's need more pressing than another's. Yet, as we saw in Chapter 1, among Bostonians, obligations toward parents and children are greatest, then come (respectively) siblings and grandchildren, in-laws, grandparents, and stepchildren, with obligations to friends often being as great as obligations to stepparents, nieces and nephews, aunts and uncle, and cousins, and obligations to ex-spouses being the least pressing. Moreover, most variations in obligations are explainable in terms of two factors: the number of intervening relatives linking ego and alter, and the relative generations of ego and alter.

Another example: Nearly any American would condemn an employer who cheats his employees, but a request to explain exactly why such an employer is despicable probably would yield ideological moralizing rather than identification of relatively straightforward principles involved in psychological processing of such an event. Yet in essence, the employer is vilified simply because he or she performs a disvalued action on a valued and relatively weak object person. The same kind of derogation falls on any actor who commits a similar act, such as a mother who beats her children, an athlete who bullies new teammates, and a doctor who abuses nurses' aides. Respondents in a survey, however, are unaware of the underlying principle that they follow, and they cannot be expected to report it.

Notwithstanding their unconscious nature, implicit cultural processes can be uncovered through social surveys. To do so, the researcher forgoes direct questioning and, instead, presents a set of situations reflecting a multiplicity of conditions, asking respondents to give a judgment in each case. Then, after the survey is concluded, the data are analyzed to reveal factors that were influencing respondents' judgments. Insights regarding obligations to kin were obtained in exactly this way, as was the insight that an actor is condemned severely for behaving in a disvalued way toward a good and relatively weak person.

Surveying Cultures: Discovering Shared Conceptions and Sentiments, By David R. Heise

Situations are presented to respondents in the form of *vignettes*, brief descriptions of circumstances or specific events that occasion the kind of judgment being investigated. Different vignettes in a study vary the conditions that are suspected of affecting judgments. Statistical analyses focus on how the conditions relate to judgments, in order to reveal the structure of respondents' unconscious normative processing. In this chapter I explicate the method of vignettes by reviewing two social science traditions that have made considerable use of the method: factorial surveys and impression-formation studies.

4.1 FACTORIAL SURVEYS

Sociologist Peter Rossi and his colleagues developed survey procedures for determining the normative influence of various factors on obligations to kin (Rossi and Rossi 1990), estimations of a family's social standing (Liker 1982; Meudell 1982; Nock 1982), estimations of household income (Shepelak and Alwin 1986), fairness of wages (Alves 1982), appropriate sentencing for convicted offenders (Berk and Rossi 1977, 1982), seriousness of child abuse (Garrett 1982), certainty about sexual harassment (Rossi and Anderson 1982), and extent of alcohol abuse (O'Brien, Rossi, and Tessler 1982).

According to Rossi and Anderson (1982, pp. 16–17), many kinds of judgments are socially structured, in that different people combine relevant characteristics in the same ways as they make the judgments. Rossi and Anderson proposed that the normative weights involved in such judgments can be estimated empirically by considering numerous cases having diverse profiles on the relevant characteristics, and by using multivariate analyses to dissect the importance of one characteristic relative to other characteristics.

What cases should be considered? They argue that real-life cases pose serious limitations.

> It would appear that "real-life" judgments would be the most appropriate data for researchers to use. ... Thus, for example, in a study of how criminal punishments are made it might appear to be relevant to examine the actual sentencing decisions of judges in a criminal court. ... However, studying such judgments presents a number of real difficulties. First, the distributions of cases and their characteristics that would come before the typical superior court are not the most useful. For example, one of the problems criminologists often pose is whether judges are sensitive to class or race in their sentencing behavior or whether they are responding more to the "legal" characteristics of the cases before them (that is, nature of evidence, credibility of witnesses, and so on). In any typical run of cases before the courts, there are very few of the critical cases that would enable one to obtain substantial amounts of empirical evidence on the issue; there are few, if any, middle-class burglars or college graduate muggers or female rapists. In short, in the real world, the various relevant characteristics of criminal cases

> tend to be correlated. [Additionally] real-world judgments . . . tend to be constrained by factors that have little to do with the objects being judged. Thus, for example, judges cannot give any sentence to convicted criminals but are guided by limits set in the criminal code. (Rossi and Anderson 1982, p. 27)

Rossi and Anderson also note that many real-life judgments are rare, such as preferences for spouses, and even selections of houses and cars are not made all that often. They conclude (Rossi and Anderson 1982, p. 28): "All of these considerations argue for moving away from the observation of actual choices or evaluations made in real-life situations to a contrived but enriched set of choices in which individuals are asked to make many judgments on sets of social objects that include combinations of characteristics that are rarely encountered in 'real-life' choice circumstances."

Rossi and Anderson refer to such stimuli as *factorial objects*; the alternative term *vignette* is used often in their research paradigm (e.g., Berk and Rossi 1982; Liker 1982; Meudell 1982; Nock 1982). In this research paradigm, vignettes—short descriptive sketches of objects or incidents—are constructed mechanically, sometimes by computers, in order to create specific combinations of characteristics. A factorial object presented by Rossi and Anderson in their study of sexual harassment provides an illustration (1982, p. 42): "Cindy M. a married graduate student often had occasion to talk to Gary T. a single 65-year-old professor. They were both at a party. She said that she enjoyed and looked forward to his class. He asked her about her other courses. He said that she could substantially improve her grade if she cooperated."

Rossi and Anderson proposed presenting vignettes to respondents for ratings on scales representing judgments of interest. They offered the following definitions (1982, pp. 28–29).

> *Dimension*—a quality of social objects or a variable characterizing such objects that can vary in kind or amount, such as sex, income, distance, and criminal actions.
>
> *Levels*—the specific values that a dimension may take. For example, "Male" is a level of the dimension "Sex," "$10,000" is a level of the dimension "income," "burglary" is a level of the dimension "criminal actions."
>
> *Object*—a unit being judged that is described in terms of a single level for every dimension. An "Object" may consist of this statement: "A male who earns $10,000 per year and who has been convicted of burglary."
>
> *Judgment*—rating, rank, or other valuation given by a respondent to an object.
>
> *Factorial object universe*—the set of all unique objects formed by all possible combinations of one level from each of the dimensions.
>
> *Factorial object sample*—an unbiased sample of the objects in a factorial object universe.
>
> *Respondent subsample*—an unbiased sample of a factorial object universe that is given to a single respondent for judgment.

In this approach, each vignette case represents a combination of one level on each dimension, and all possible cases constitute the factorial object universe. Dimensions in the object universe are uncorrelated (orthogonal) because every level of each dimension is combined with every level of other dimensions. Moreover, the distribution of objects is rectangular on every dimension, since each level is represented by the same number of constructed objects. Rossi and Anderson (1982, p. 30) noted the advantages that this provides in uncovering the implicit structuring of judgments, and they submitted that these advantages also hold for random samples from the object universe.

> It is the approximate orthogonality among dimensions that makes it possible to disentangle the separate effects upon judgments of dimensions that are ordinarily (in the real-world context) correlated. Rectangularity allows observations to be made along all segments of a dimension, such that the response of judgments to variations in a dimension can be estimated more efficiently as to both size and form.
>
> A "factorial object sample" is an unbiased sample of the members of a "factorial object universe" that preserves the essential characteristics of the object population—namely, that the correlations among dimensions asymptotically approach zero as the size of the sample increases and that the distribution of objects along any of the dimensions tends asymptotically toward rectangular. We will argue that because such samples have the same properties (asymptotically) as the populations from which they were drawn, analyses of such samples will result in estimates that asymptotically converge on population parameters. The application of sampling makes it possible to handle many dimensions and many levels.

A factorial design can generate huge numbers of cases in the factorial object universe when there are numerous dimensions and many levels of each dimension. The total number is calculated by multiplying the numbers of levels in each relevant dimension. For example, to study status judgments of families, there might be 10 levels of income; 50 head-of-household occupations and 50 occupations for spouses; two races, white and black; and 10 levels of family size: resulting in a universe size of 500,000, many more than any one respondent could rate. The huge size of some factorial object universes means that traditional factorial designs from experimental research cannot always be used for studying judgments.

Rossi and Anderson (1982, p. 30) proposed that instead of having each respondent rate all objects in the object universe, a respondent could rate a random sample from the universe, and the ratings of multiple respondents could be pooled. Any one respondent thus rates a relatively small subsample of factorial objects, and all respondents together generate a relatively large sample of the factorial object population. A factorial survey consequently involves two separate sampling procedures. One sampling procedure draws from the object universe in order to generate multiple sets of vignettes for respondents to rate. The other sampling procedure draws from the human population of interest in order to assemble a set of respondents. Pooling the M judgments made by each of N respondents yields another simple random sample from the universe, but now of size $R = NM$ (Rossi and Anderson 1982, p. 32).

Ordinarily, even a pooled sample is much smaller than the object universe. For example, the study of family statuses mentioned above might be fielded with 300 respondents, each of whom rates 30 vignettes, giving a factorial object sample of 9,000 rated vignettes, substantially smaller than the 500,000 hypothetical vignettes in the object universe. In situations like this, with large object universes and fairly modest-sized samples from the object universe, it is unlikely that any two respondents will rate the same object, so each vignette in the universe will be rated either by no one or by one respondent, and only unusually by more than one respondent.

Each level of a dimension in the object universe contributes to ratings of multiple vignettes: about Q times, where $Q = R/L$, with L being the number of levels in that dimension. For example, in the study of family statuses, vignettes with white families would get rated about 4,500 times, and about 4,500 vignettes involving black families would be rated; meanwhile, heads of household who are registered nurses (one of the 50 occupations) would get rated in about 180 vignettes.

Data from factorial surveys typically are analyzed in one or more of three modes, as detailed below. All three modes typically employ the assumption that dimensions do not interact in generating responses to objects. That is, the causal model relating object dimensions to judgments is assumed to be strictly linear. Effects of various characteristics therefore add up neatly without taking into account which characteristics are conjoined in particular cases. For example, in the study of family statuses, the assumption implies that the status of black families with registered nurses as heads of household can be determined by combining the typical status of black families with the typical status of families in which the head of household is a registered nurse; no adjustment should be needed to account for a family being both black and headed by a registered nurse.

Data to check some interactions of this type typically would exist in the factorial object sample. For example, the illustrative study outlined above would have ratings for around 90 vignettes presenting black families with registered nurses as heads of household, so one could check whether the assumption of linear combination of effects actually holds for this conjunction. However, the sparsity of the sample relative to the universe of objects interferes with examining very high order interactions. For example, chances are that the data contain no rating for a large black family in which the head of household is a registered nurse and the spouse is unemployed; the probability of this conjunction occurring in the 9,000 cases of the factorial object sample is about 0.02.

Data from a factorial survey are not as complete as data from a factorial experiment, because the factorial universe is sampled rather than assessed exhaustively, and therefore one cannot assay interaction effects involving all or most of the factorial dimensions. This is a matter of economy rather than necessity. In principle, even if not practically, larger samples of respondents could be drawn, and respondents could be given larger samples of vignettes to judge, until one or more judgments for every object in the universe became available, allowing all possible interactions among dimensions to be studied.

4.1.1 Means Across Vignettes

The fact that objects characterized by a particular level of a dimension get rated repeatedly in a factorial survey leads to one of the standard modes of analyzing data from such surveys: computing the mean ratings of vignettes at each level of each dimension. The means of levels are essentially independent of one another, due to the orthogonality and rectangularity issues mentioned above, so the means can be compared within dimensions and across dimensions to assess quantitatively how various characteristics of objects contribute to judgments.

As an illustration, consider some hypothetical means of family status ratings on a 10-point scale: white families, 5.0; black families, 4.7; families whose heads of household are registered nurses, 5.5; families whose heads of household are unemployed, 4.5. Such numbers would indicate that race is a factor in status judgments, with black families being rated lower than white families; that occupation is also a factor; and that occupation is more important than race, since the difference in means between a head of household who has a respected job and one who is unemployed is larger than the difference in means between white families and black families.

Means of levels within dimensions can be interpreted as indirect evaluations of the various characteristics distinguishing the objects. For example, the hypothetical means above provide a status measurement for white families and a status measurement for black families. Rossi and Anderson (1982, p. 63) noted that interpreting means in this way resonates with the functional measurement approach promoted by psychologist Norman Anderson (e.g., Anderson 2008).

Examples of this analytic approach are provided by Garrett (1982) in her study of child abuse and by Rossi and Rossi (1990) in their study of kinship obligations, which I reviewed at length in Chapter 1.

4.1.2 Consensus Model

Estimating regression models constitutes the standard method of dealing with factorial survey data. Two major variations were presented by Rossi and Anderson. The first approach assumes that a single model applies to the information processing of all respondents, corresponding to normative consensus regarding how judgments are formed from object characteristics. Regression analyses estimate parameters of such models, and according to Rossi and Anderson (1982, p. 21), the squared multiple regression coefficient, or coefficient of determination, provides a basis for appraising the extent of normative consensus.

The model in this case stipulates that respondent *i*'s judgment regarding vignette *m* within the respondent's set of vignettes is determined by the sum of a constant *c*; for each of the *D* vignette dimensions, a dimension coefficient multiplied by the vignette's interval-scale value *V* on that dimension; and an error term *e* for that respondent judging that vignette:

$$J_{mi} = c + \sum_{d=1}^{D} b_d V_d + e_{mi} \tag{4.1}$$

Pooling the *M* responses of the *N* respondents amounts to assuming that the dual indexes *m* and *i* can be treated as a single index *k* extending from 1 to *NM*:

$$J_k = c + \sum_{d=1}^{D} b_d V_d + e_k \tag{4.2}$$

This model can be estimated by regressing all judgment ratings in the pooled sample on the set of dimension values of the vignettes being judged. The result is a large number of judgments to study, as if an individual respondent had worked at the judgment task for many hours.

Dimensional values for a vignette often are unavailable on an interval scale, in which case a qualitative value has to be assigned, using sets of dummy variables: one zero–one variable for each level of the dimension up to the number of levels on that dimension L_d, minus one. When all dimensions are measured in terms of qualitative levels, the model becomes

$$J_{mi} = c + \sum_{d=1}^{D} \left(\sum_{l=1}^{L_d - 1} b_{l_d} V_{l_d} \right) + e_{mi} \tag{4.3}$$

V_{l_d} in this model is either a zero or a one, depending on whether or not the vignette has the value of level *l* on dimension *d*.

Pooling the *M* responses of the *N* respondents allows the qualitative version of the model to be estimated by regressing judgment ratings on the *D* sets of dummy variables. A dimension's overall influence on judgments could be summarized by computing a sheaf coefficient (Heise 1972) for that dimension's set of dummy variables. Dimensions measured on interval scales can be combined with dimensions implemented with dummy variables, although I forgo presenting a mathematical representation of this. Introductory books on regression analysis discuss the procedures required.

Results based on the consensus regression model as outlined above were reported in all but one of the chapters in Rossi and Nock's (1982) collection of factorial survey studies. Coefficients of determination in these studies range from 0.14 to 0.50. The relative importance of the error term in regression models is the proportion of residual variance, calculated as one minus the coefficient of determination. This quantity ranges from 0.50 to 0.86 in these studies, indicating that half or more of the variance in judgments in each study comes from the error term, e_{mi}. The assumption of consensus in judgments seems precarious if these figures are used as supporting evidence.

Yet in factorial surveys, proportions of residual variance compound multiple kinds of error. The error term's i subscript relates to differences among respondents in how they rate vignettes; these individual differences include real disagreements in judgments, and it is in that sense that the error term and computed proportions of residual variance provide a measure of normativeness. However, as discussed in Chapter 5, respondent errors in ratings have multiple components, such as respondents' individual differences in scale usage, respondents' transient variations in outlook, and respondents' differing interpretations of words. The error term in equation (4.1) or (4.3) also includes deviations associated with respondents' erroneous estimations of general norms (Rossi and Berk 1985, p. 340). Computed proportions of residual variance are affected by all such factors, and thereby they are adulterated as indicators of normative consensus.

Moreover, the error terms in equations (4.1) and (4.3) also have an m subscript, indicating that a model may erroneously predict judgments about vignette m, even if there were no respondent errors at all. This kind of error occurs when key dimensions have been omitted from the model, or when interactions among dimensions undermine the linear representation of effects from each dimension. The proportion of residual variance computed in factorial surveys aggregates this modeling error with respondent errors, whereupon the proportion provides little practical information about either the adequacy of the respondents or the adequacy of the model.

4.1.3 Averaging Individual Ratings

A variation of regression analysis in factorial surveys employs mean ratings of vignettes as the judgments to be explained. This reduces the impact of individual differences among respondents by transforming the standard deviation of ratings for a vignette into the standard error of the mean rating for that vignette: a smaller quantity since the standard error is the standard deviation divided by the square root of the number of averaged ratings. As means are based on greater number of ratings, respondent errors become more and more negligible, leaving just the deviations of particular vignettes from the predictions made by the model. Thus, when mean ratings are the dependent variable, the computed proportion of residual variance can be a reasonably good measure of the adequacy of the model.

Some factorial surveys do report regressions employing mean ratings rather than the ratings of individual respondents. Mean ratings were analyzed in the study of appropriate sentencing for convicted offenders (Berk and Rossi 1982) and in the study of kinship obligations (Rossi and Rossi 1990). The sentencing study, with means based on three to 12 respondents, obtained a coefficient of determination of 0.70 for ratings of desired treatments of criminals, and correspondingly, the computed proportion of residual variance equaled 0.30. Regressions to describe structuring of kinship obligations, which used judgment means computed over about 496 judgments on the average, obtained a coefficient of determination of 0.94 for judgments about obligations to provide comfort and for judgments about giving money, and the corresponding proportion of residual variance was 0.06. These two instances demonstrate that averaging the dependent variable over aggregated cases increases squared multiple regression coefficients dramatically and reduces proportions of residual variance. In turn, this reveals that linear models for predicting judgments from characteristics of vignette objects can be quite powerful.

The predictor variables in factorial surveys—the various dimensions characterizing a vignette object—ordinarily are assessed without any notable error, and consequently, estimates of regression coefficients are unbiased, regardless of the amount of random error in judgments, as noted by Berk and Rossi (1982, p. 159). Thus, the studies mentioned above, in which coefficients of determination were low, probably provided accurate estimates of model parameters, assuming that raters were in reasonable consensus about how object characteristics translate into judgments.

4.1.4 Variant Norms

Within a general population, some subgroups of respondents may integrate information about an object's characteristics into a judgment in ways that differ systematically from the general norms. The subgroup patterns can be investigated with variants of the regression models presented above.

Rossi and Anderson (1982, pp. 23–24) advocated elaboration of regression models to assess biases (thresholds, in their terminology) in judgments associated with respondent attributes such as age, gender, or education. Such biases can be intrinsically interesting, as, for example, when Rossi and Anderson found that, generally, higher levels of sexual harassment are perceived by females than by males. Moreover, making such biases explicit removes them from the error term of the equation, so statistical tests are more powerful.

This kind of elaboration turns equation (4.1) into

$$J_{mi} = c + \sum_{d=1}^{D} b_d V_d + \sum_{a=1}^{A} b_a Z_{ai} + e_{mi} \qquad (4.4)$$

In this equation, Z_{ai} is respondent i's value on respondent attribute a (either an interval-scale measurement or a zero–one dummy variable) and A is the total number of respondent attributes under consideration. Equation (4.3) can be elaborated similarly, and either of these elaborations can be extended to incorporate sets of dummy variables for qualitative respondent attributes. The models are estimated by pooling the M responses of the N respondents, and regressing values of judgments on values of vignette variables along with values of respondent variables.

A variation of this kind of model uses zero–one variables signifying respondents' individual identities as the relevant attributes. The b_a coefficients then indicate how the constant c is to be adjusted for each respondent in order to allow for that person's bias when making judgments of the kind being considered. A domain is in "relative consensus" if respondents differ only in the biases (thresholds) of their judgments (Rossi and Berk 1985, p. 339).

A more general approach allows not only for subgroup biases but also for subgroup adjustments in the structural parameters that define how information is integrated:

$$J_{mi} = c + \sum_{d=1}^{D} b_d V_d + \sum_{a=1}^{A} Z_{ai}\left(\sum_{d=0}^{D} b_{ad} V_d\right) + e_{mi} \tag{4.5}$$

In this representation, V_0, the value of dimension zero, should be understood as the number 1. The structural parameters, b_d, indicate the normative influence assigned to vignette dimension d when converting the value on that dimension, V_d, into a judgment. The additional structural parameters, b_{ad}, indicate how the influence of each vignette dimension is to be adjusted to take account of respondent i's value on individual characteristic a; these parameters are the regression coefficients for product variables $Z_{ai}V_d$. The coefficients b_{a0} correspond to the b_a coefficients in equation (4.4), and they indicate how respondent i's value on characteristic a affects the constant, c. Such a model is estimated by regressing the NM judgments in the pooled data set on all variables in the model.

Equation (4.5) could be elaborated to allow for qualitative variations in vignette dimensions and for qualitative variations in individual attributes. The mathematical representation is bristly and I forgo presenting it here, but the general idea is straightforward. Regress judgments on the sets of dummy variables representing the vignette dimensions (corresponding to V_d), on the sets of dummy variables representing respondent attributes (in order to estimate the b_{r0}), and on the arrays of dummy variables formed by crossing each set of vignette dummy variables with each set of respondent dummy variables (corresponding to $Z_{ri}V_d$, where $d > 0$). Setting up such an analysis, and interpreting the results, is challenging when there are multiple vignette dimensions and multiple respondent attributes, but is not intrinsically problematic when one proceeds systematically.

4.1.5 Homogeneity

The factorial survey approach developed within the framework of traditional surveys, so the presumption almost always was that respondents should be acquired via probability sampling of standard populations. Yet the focus of the approach is on norms—cultural matters—so analyses ideally focus on culturally homogeneous respondents rather than on diverse respondents from a general population. The heterogeneity problem posed by probability sampling led to the development of ways to extricate homogeneous groups from sample diversity. Identifying subgroup thresholds by estimating equation (4.5) is one such expedient. Additionally, Rossi and Anderson (1982, pp. 25–27) considered several kinds of analyses exploiting the multiple judgments of vignette objects provided by each respondent in order to partition respondents into groups of individuals who homogeneously employ the same principles for translating characteristics of vignette objects into a particular kind of judgment.

Following a major theme in this book, the heterogeneity problem in factorial surveys actually should be addressed directly by relinquishing random samples of respondents from general populations. Instead, respondents should be selected from the actors who reproduce culture by making judgments of interest on a regular basis, with special consideration given to those with the most experience and expertise. Dealing with persons who are not ideal in this sense wastes resources, and contaminates data with errors of judgment by those who lack inculcation into relevant cultural norms.

Berk and Rossi's study of sentencing of convicted offenders did something very close to this. "The overall purpose ... was to gauge the receptivity to reforms in adult corrections among persons who were potentially instrumental in initiating, approving, and carrying out programs within state corrections systems. It was to be, therefore, a study of 'elites'" (Berk and Rossi 1977, p. 3). Berk and Rossi decided on specific numbers of respondents to be acquired among official decision-makers such as governors, heads of correction agencies, and members of corrections committees in legislatures among law-enforcement personnel such as wardens, police chiefs, judges, and prosecuting attorneys, and among other partisans, such as mayors and officials of involved organizations such as the ACLU and police benevolent associations (Berk and Rossi 1977, Table 1-1). Although they covered the ideological spectrum, these expert respondents were found to agree on facts regarding their correctional system (1977, p. 138), even if not on what changes would be desirable. Similarly, Rossi and Berk (1985, p. 341) noted that in a Chicago study of the seriousness of crimes, judges, probation officers, and prosecuting attorneys had lower error variances than respondents from the general population, and their ratings corresponded better with mean ratings across several studies.

Rossi and Rossi's (1990) study of kinship obligations also met the goal of acquiring respondents who actively reproduce culture by making the relevant judgments on a regular basis, as pointed out in Chapter 1. Some variants in family culture no doubt exist in Boston, but the variants were definable by application of the equations given above. Meanwhile, the notion of a generally shared family culture in the region is credible within the context of the dramatic variations in kinship structure recorded by anthropologists (e.g., Murdock 1949).

4.1.6 Summary

Factorial surveys use vignettes to investigate processes that are influenced by multiple components, where each component has divisions (often, numerous divisions) that combine with divisions of other components to yield outcomes of the process. The number of combinations of divisions of components typically is so large that there is no hope of including all combinations in a study, in the manner of analysis of variance studies. Rather, combinations are randomly sampled. Each sampled combination is used to create a vignette that is presented to respondents in order to elicit their judgments. The pooled judgments of many judges dealing with manifold combinations are then correlated with the components involved in the combinations, resulting in an assessment of which components influence the process, and to what extent.

The approach involves a fundamental assumption that each component acts independent of the others. Although a few lower-order interactions among components might be estimable in a particular study, the strategy of sampling combinations of divisions of components precludes estimation of component interactions in general.

Various analytic procedures have been specified to examine differences in processing within different groups and to identify relatively homogeneous groups of people who arrive at judgments in similar ways. However, I argued that it is best to recognize the cultural nature of such studies and to select respondents from culturally homogeneous groups. Several factorial surveys actually have done this.

4.2 IMPRESSIONS FROM EVENTS

Impression formation occurs when concepts combine in the mind, generating emergent feelings about the constituent concepts. One kind of impression formation arises from discerning events: The dynamic combination of an actor, a behavior, and an object person produces new feelings about all three elements. Another kind of impression formation involves noticing that a person in some identity has some kind of attribute; the combination of identity and attribute produces a new feeling about the person.

Impressions generated by interpersonal events are studied by presenting verbal descriptions of events (vignettes) to respondents, and asking the respondents to report their feelings about different aspects of the events.

Psychologist Harry Gollob pioneered impression-formation research with vignettes describing interpersonal occurrences (1968, p. 341). He described his study as focusing "on the problem of predicting the evaluative rating (i.e., the good–bad rating) of a sentence subject as it is described by the total sentence. In order to reduce the problem to manageable proportions, the study deals only with sentences which fit the sentence frame: The *adjective* man *verbs noun* (e.g., 'The vicious man likes beggars.' 'The kind man praises communists.') Thus, from characteristics of the adjective, verb, and object we wish to predict the evaluative rating of the man described by the sentence." The relevant characteristics of the adjective, verb, and object that Gollob considered were their evaluations when rated in isolation.

Gollob (1968) showed that a fairly simple regression equation,

$$S_e = -0.43 + 0.39A_e + 0.48V_e + 0.15V_eO_e \tag{4.6}$$

accounted for 86 percent of the variance in respondents' evaluations of the man in the various sentences that he presented to respondents. [In this equation, S_e stands for the predicted rating of the subject (the man), A_e stands for the rating of the adjective in the sentence, V_e stands for the rating of the verb, and O_e stands for the rating of the object person.] Generally speaking, the man was described as good if characterized by a positive adjective (A_e), if engaging in a good action (V_e), and if the action befitted the character of the object person (V_eO_e), good acts being directed at good objects and bad acts at bad objects.

A similar study (Gollob and Rossman 1973) examined impressions of an actor's "power and ability to influence others." In this case, the sentences were of the form[1] "Bill helped the corrupt senator" and "Joe threatened the incompetent man." Predictions of the actor's potency were made from evaluation and potency ratings of the verb and of the object person. Again, a relatively simple regression equation,

$$S_p = -0.32 + 0.77V_p + 0.21V_eO_e - 0.09V_eO_p \tag{4.7}$$

accounted for a large percentage of the variance in ratings of actor potency—74 percent. In this equation S_p is the predicted potency of the subjects of the sentences, V_p is the rated potency of someone directing the given action at a man, and O_p is the rated potency of the object person. Thus, the impression of an actor's potency reflects the potency of the act (V_p), the evaluative consistency between the act and the object person (V_eO_e, which Gollob interprets as an indicator of how justly the actor behaves), and congruency between the evaluation of the act and the potency of the object person. The congruency

[1]Gollob and Rossman's use of personal names makes their stimuli more readable, but personal names do contribute additional affective associations to the stimuli (Lawson 1973; Lawson and Roeder 1986).

term, including its negative sign, suggests that an actor is seen as merciful when good acts are directed at weak objects, courageous when bad acts are directed at strong objects, sycophantic when good acts are directed at strong objects, and base when bad acts are directed at powerless objects.

My colleagues and I expanded Gollob's work to all three dimensions of affect, to various kinds of events and to cultures other than American (Britt and Heise 1992; Heise 1969a, 1970a, 1979; Heise and Smith-Lovin 1981; MacKinnon 1994; Smith 2002; Smith and Francis 2005; Smith, Matsuno, and Umino 1994; Smith-Lovin 1979, 1987b; Smith-Lovin and Heise 1982). All of these studies predicted the *mean* Evaluation, Potency, or Activity of an event element from *mean* EPA ratings of the same elements outside the context of the event, thus attaining the benefits discussed in Section 4.1.3.

Most of the work focused on interpersonal events consisting of an actor behaving toward an object person. A description of an event combined nouns designating actor (A) and object (O) identities, with a verb designating the interpersonal behavior (B). For example, the identities of mother and daughter and the behavior of kissing could be used to create the ABO combinations of "a mother is kissing her daughter" and "a daughter is kissing her mother."

New impressions about an event's actor, behavior, and object form as the event combines feelings about the three elements. Consider, for example, how the impression of a mother changes in the following events.

A mother is kissing her daughter.
A mother is hitting her daughter.
A mother is hitting a bully.

In terms of the EPA dimensions, the mother in the first event is extremely good, quite potent, and slightly active; the mother in the second event is quite bad, potent, and active; and the mother in the third event is slightly good, potent, and active.

Impressions created by an event are predictable from feelings about event constituents before the event occurs. For example, the evaluation of the mother in the three events above derives largely from general feelings about the behaviors of kissing and hitting and from the way those feelings combine with feelings about the object persons. Good behavior creates a good impression of the actor; bad behavior directed toward a good person creates a bad impression of the actor; and bad behavior directed toward a bad person creates a slightly good impression of the actor.

Regression equations can be obtained to predict respondents' outcome impressions of event constituents from respondents' assessments of event constituents when presented outside the context of any event. The regression equations include terms for main effects and for interactions. To illustrate, the following is a simplified equation for predicting the outcome Evaluation of the actor in an event (A'_e) from the pre-event evaluations of the actor (A_e)

and behavior (B_e), and from the product of the behavior and object person evaluations (B_eO_e), as in the parallel equation (4.6).

$$A'_e = k + b_1A_e + b_2B_e + b_3B_eO_e \tag{4.8}$$

The constant (k) and the weights (b_1, b_2, and b_3) are estimated via regression analysis. Impression-formation studies of this kind typically estimate nine equations: the outcome Evaluation, the outcome Potency, and the outcome Activity of the event's actor, behavior, and object person. Some studies also have estimated the outcome Evaluation, Potency, and Activity of the setting in which the event occurred.

My early impression-formation studies (Heise 1969a, 1970a, 1978) employed 24 to 69 event sentences as the bases for regression analyses. Such small numbers limited the precision of the parameter estimates. Nevertheless, the studies did replicate the general features of Gollob's (1968, 1973, 1974b) work: that is, the outcome impression of an actor is built from pre-event feelings about the actor, behavior, and object, with pre-event feelings having both direct impacts and effects produced by one event element interacting with another. Additionally my early studies showed that impressions form on all three dimensions of affect—Evaluation, Potency, and Activity—and that an event produces impressions of the action and impressions of the object person in the event as well as impressions of the actor.

4.2.1 UNC Study

A project sponsored by the National Institute of Mental Health in 1978 and 1979 and conducted at the University of North Carolina (UNC) expanded the database to 515 event sentences, so my students, colleagues, and I were able to obtain greatly improved estimations of coefficients in impression-formation equations. The UNC project was reported in a special issue of the *Journal of Mathematical Sociology* (1987, Volume 13, Numbers 1 and 2), which was reprinted as a book (Smith-Lovin and Heise 1988). I summarize the essential features of the project's main study (Smith-Lovin 1987b) as an example of how vignettes are constructed in impression-formation research.

A person's identity, or an interpersonal behavior, can vary in eight major ways along the EPA dimensions, as shown in Table 4.1. Ideally, we would gauge impression formation effects from a set of events in which every kind of actor is crossed with every kind of behavior and every kind of object person. The following examples illustrate some possible event descriptions in such a study:

The champion entertained the friend. (A+++, B+++, O+++)
The loafer laughed at the champion. (A−−−, B−−+, O+++)
The grandparent beat up the child. (A++−, B−++, O+−+)

TABLE 4.1 EPA Patterns for Identities and Behaviors

Pattern	Sample Identities	Sample Behaviors
Good, potent, active (E+P+A+)	a friend, a champion	to entertain
Good, potent, passive (E+P+A−)	a grandparent, a priest	to console
Good, impotent, active (E+P−A+)	a child, a teenager	to request something from
Good, impotent, passive (E+P−A−)	a librarian, a senior citizen	to observe
Bad, potent, active (E−P+A+)	a bully, a gangster	to beat up
Bad, potent, passive (E−P+A−)	an executioner, a serial murderer	to condemn
Bad, impotent, active (E−P−A+)	a prostitute, a criminal	to laugh at
Bad, impotent, passive (E−P−A−)	a loafer, a dropout	to beg

The fully crossed design assures that a great range of different kinds of events is represented in the study. Moreover, such a design increases the efficiency of regression analyses by minimizing correlations among the measurements used to predict outcomes (i.e., the EPA values of actors, behaviors, and objects).

Fully crossing all eight kinds of actors with all eight kinds of behaviors and with all eight kinds of object persons yields a total of 8·8·8 = 512 event sentences. Event sentences in the UNC study were constructed in this way. The study incorporated one event sentence for each of the 512 actor–behavior–object combinations, plus three event sentences with neutral elements, yielding a total of 515 event sentences.

Each event sentence was presented three times in the study, each time to different respondents. In one presentation the actor in the event was underlined (e.g., The *loafer* laughed at the champion), in another presentation the behavior was underlined (e.g., The loafer *laughed at* the champion), and in the third presentation the object person was underlined (e.g., The loafer laughed at the *champion*). Respondents were instructed to rate how they felt about the underlined person, or action, in the context of the event. Three bipolar scales followed each stimulus, anchored as follows:

1. *Evaluation:* good, nice versus bad, awful
2. *Potency:* big, powerful versus little, powerless
3. *Activity:* fast, young, noisy versus slow, old, quiet

Besides rating identities and behaviors in the context of events, the study design required out-of-context ratings of all the identities and behaviors used in the event sentence vignettes: for example, a champion, a friend, a loafer, a grandparent, a child, plus entertaining someone, laughing at someone, and beating up someone. Additionally, as mentioned in Chapter 3, where this study was first mentioned, sentiment measurements were obtained for 721 identities, 600 behaviors, 440 modifiers, and 345 settings. Beyond this, EPA ratings also were obtained for a modifier-identity study (Averett and Heise 1987), and likelihood ratings of each of the 515 events were obtained for a study of event normativeness (Heise and MacKinnon 1987). "There were 40 different forms of the questionnaire, containing 1069 distinct pages (for a total of 5345 stimuli)" (Smith-Lovin 1987b, p. 42). The 40 forms were completed by different subgroups within the total of 1,225 respondents.

The UNC study resulted in a set of nine equations for predicting the outcome Evaluation, Potency, and Activity of an event's actor, behavior, and object. The study indicated that impression formation processes are quite complex, with nine first-order terms, 14 two-variable interactions, and five three-variable interactions appearing in one or more of the nine equations. Smith-Lovin (1987b) presented a detailed description of the nine equations, both quantitatively[2] and verbally, and I have interpreted various terms in the equations elsewhere (Heise 2007, Chapter 6), so I offer no detailed account of the equations here. However, it is worth noting that the new equations verified the importance of effects previously found to be important in impression formation, by Gollob (1968, 1973) and myself (Heise 1969a, 1970a, 1978).

4.2.2 Later Studies

The next impression-formation study was conducted in Japan by Herman Smith, Takanori Matsuno, and Michio Umino (1994). In Chapter 3 I related details regarding the Japanese rating scales, concepts presented for sentiment assessments, and respondents. Here I supplement that information with a brief description of the Japanese vignette study.

Forgoing a complete Japanese replication of the UNC study because of the costs of dealing with more than 500 event sentences, Smith, Matsuno, and Umino took a sampling approach similar to that promoted by Peter Rossi, as discussed in the first part of this chapter. They classified the Japanese identities and behaviors into the types listed in Table 4.1.

> From these elements we created a nearly balanced factorial set of randomly formed Japanese events. Then we used a computer program to randomize identities and behaviors into ABO sentences. Thus, rather than sentences with subjects, actions, and object-persons in sentence-type agreement, we produced randomized sentences. The main purposes of factorial design are (1) to ensure lack of correlation between treatment variables and (2) to present subjects with randomized and balanced sets of stimuli. Our modification of the Smith-Lovin design accomplishes these goals. . . . Two randomly created pools of 50 sentences each served as stimuli. (Smith, Matsuno, and Umino 1994, p. 131)

[2]Smith-Lovin (1987b) presented equations obtained with LISREL, a full-information estimation method. Later I reestimated the equations using ordinary least-squares (OLS) (Heise 1991), and the OLS results were incorporated into program *Interact* (Heise 1997) because they seemed to provide better outcomes in simulations.

The Japanese impression-formation equations turned out to be similar to American equations in important respects: most of the effects found by Gollob (1968, 1973) and by me in my early studies (Heise 1969a, 1970a, 1978) showed up in the Japanese equations. However, the Japanese equations were more succinct than the equations derived in the UNC study, with fewer three-way interactions.

Neil MacKinnon estimated impression-formation equations in Canada. Chapter 3 provides details concerning the Canadian rating scales, the concepts rated, and the respondents. The additional issue here is the design of the vignette study, which MacKinnon (1985, p. 20) described as follows.

> The American study used a Latin Square design for which 8 × 8 × 8 = 512 + 3 additional sentences = 515 events were constructed. While the Canadian study employed the same design, event sentences were written only for those cells corresponding to interaction terms found significant in the American study and for which there was any judged likelihood of significance occurring in any replication of the American study. Thus, the Canadian study employed only 214 event sentences, as compared to 515 in the American study. In addition, while the format (the particular ABO configuration and EPA profile) for each event sentence was preserved in the Canadian study, the content differed because a new set of sentences was written for the Canadian study to correspond to the major institutional areas (family, legal, etc.) scanned by the study.

MacKinnon found that the structure of the Canadian equations largely replicated the structure found in the American UNC study, although the Canadian equations were somewhat more complex, with a greater number of two-way interaction terms.

Tobias Schröder's 2007 study in Germany—conducted via the *Surveyor* program, with scales and respondents as described in Chapter 3—included a component devoted to estimating German impression-formation equations.[3] Schröder presented respondents with 100 event sentences, his design being a replication of the design developed by Smith, Matsuno, and Umino (1994) for the Japanese impression-formation study. Schröder replicated the EPA configuration of the Japanese sentences rather than simply translating the Japanese sentences to German. For example, the Japanese event "A rival plays with the rascal" was turned into "An athlete saves a lout" in order to maintain the same EPA configuration (A+++B+++O−++).

In the German study, each event was presented to respondents as a discrete sentence. The event element to be rated within the context of the event was identified by a subsequent question. For example, one stimulus (translated to English) was: "An athlete saves a lout. How do you feel about the lout in this situation?"

[3]My discussion of Schröder's impression-formation study is based on an incomplete manuscript that he kindly provided for me in 2009.

The Schröder study of impression formation in Germany found some of the effects that had emerged in previous studies of impression formation. However, the German study recorded some substantial divergences from impression formation processes found in the United States, Canada, and Japan. For instance, Germans' Evaluation impressions of an actor were heavily dependent on pre-event feelings about the actor and the behavior, as in previous studies. However, the German study found actor Evaluation benefiting only a little from a just-world effect (B_eO_e), and not at all from congruency of behavior evaluation with object potency (B_eO_p).

4.2.3 Self-Directed Action

Two vignette studies examined impressions created by self-directed actions such as praising oneself or medicating oneself. Such events have only two elements: the actor's identity and the behavior directed toward the self. The actor implicitly is the object of the action as well as the actor.

Britt and Heise (1992) presented 256 vignettes, along with identities and behaviors that were used to construct the vignettes, to 409 American undergraduates. The stimuli were divided into five sets, with each stimulus set being given to 30 or more males and 30 or more females. The respondents rated the Evaluation, Potency, and Activity of actors and behaviors out of context and also in the context of the vignettes, using the *Attitude* data-gathering program. (The *Attitude* program and its scales are discussed in Chapter 2.) "Event descriptions were presented in a frame intended to enhance respondents' visual imagery; for example, 'You see the champion delighting herself. She seems' or 'You see the gangster forgiving himself. The act seems'" (Britt and Heise 1992, p. 337).

Vignettes were constructed from a design that crossed all eight EPA patterns of actor identities with all eight EPA patterns of behaviors. The resulting 64 event configurations were replicated four times, half with male pronouns referencing the actor and half with female pronouns referencing the actor.

Britt and Heise (1992) found that impressions from self-directed actions formed in ways that differed substantially from the process of forming impressions in interpersonal events. In particular, negative equation constants indicated that in general, self-directed actions made an actor seem less good, potent, and active; and small coefficients for behavior evaluation and potency indicated that self-directed behaviors could do little to enhance a person's presentation of self as good and potent.

The Britt and Heise (1992) study was the first to consider whether the gender of the actor in an event makes a difference in impression-formation processes. They found one difference: Females more than males were expected to behave in a manner consistent with their identity, acting more positively toward the self than males when in valued identities, and more negatively when in disvalued identities.

Britt and Heise also ran a small auxiliary study to determine if respondents processed events the same way when they themselves were the actors, as opposed to observing others as actors. This was done with vignettes of the form, "Imagine you're a freshman pampering yourself. You feel." They found that impressions of self in action depended mainly on the behavior, with relatively little effect from the self's identity, and no evaluative consistency effect (Britt and Heise 1992, p. 345).

Smith and Francis (2005) reported a Japanese replication of parts of the Britt and Heise study. They collected ratings from 240 students in the greater Tokyo area, divided equally by gender, using the Japanese version of the *Attitude* data collection program (the program and its scales are discussed in Chapter 2). Two vignettes were created for each of the 64 actor–behavior configurations generated by crossing each EPA type of actor with each EPA type of behavior (the eight EPA types are given in Table 4.1). "A translation of one of the 128 sentences created is: 'You observe a fashion designer delighting herself. She seems (The behavior seems) ___ to you" (Smith and Francis 2005, p. 824).

Equations reported by Smith and Francis (2005) had smaller negative constants than the American study, suggesting that in Japan self-directed actions are less tarnishing of feelings about the actor and behavior. Other differences in Japanese and American equations indicated that the Japanese are less influenced than Americans by characteristics of the actor, when developing impressions from self-directed actions.

4.2.4 Qualified Behaviors

Neil MacKinnon (1985) modified the usual vignettes for studying impression formation and event likelihoods by adding frequency adverbs to behaviors. For example, his study included the following descriptions.

The psychotic never admires the psychiatrist.
The nurse rarely quarrels with the doctor.
The physician occasionally avoids the intern.
The maniac frequently attacks the paranoid.
The tutor always idolizes the disciplinarian.

Analyses of the perceived likelihood of the circumstances described revealed an inverted-U relation between the adverbs and the average perceived probabilities of happenings described with the adverbs. Circumstances described with the "occasionally" modifier had the highest average likelihood. Circumstances described with the "rarely" and "frequently" modifiers had about the same average likelihood, which was higher than the average likelihood of circumstances whose behavior was described as never occurring or always occurring. Thus, an occasional occurrence of most events seemed most plausible to respondents. At the same time, respondents' ratings of likelihood also were influenced by the degree to which a circumstance disturbed their feelings about the people and action involved, as discussed in Section 4.4.

MacKinnon (1985) estimated impression-formation equations for circumstances specified with frequency adverbs. The overall structure of the equations was largely the same as obtained for specific events, but interactions involving the object person seemed more important for ongoing circumstances than for specific events.

Lisa Slattery Rashotte found that demeanors exist for every Evaluation–Potency–Activity configuration, as in the following examples (Rashotte 2001, Table 1): +++ laughing, ++– stretching, +–+ blinking, +–– tilting the head, –++ making a fist, –+– putting hands on hips, ––+ rolling one's eyes, ––– sucking one's fingers.

One study (Rashotte 2002) sought to predict the impression of demeanor combined with a transitive social action (e.g., smiling and comforting someone). Evaluations of such combinations essentially averaged the goodness of the demeanor and the action, with an additional large effect from evaluative consistency between the demeanor and the action. Potency of the combination was affected most by the potency of the demeanor; potency of the transitive action actually detracted from potency of the combination; and potency of the combination also accrued from evaluations of both demeanor and action, activity of the action, and evaluative consistency of demeanor and action. The activity of the combination more or less averaged the activities of the two components, with an additional positive effect from the demeanor's potency and from the action's goodness and weakness.

A study of 64 events in a Graeco–Latin square design (Rashotte 2001) added descriptions of an actor's demeanor to typical vignettes of impression-formation studies. Rashotte found that impressions of an event's action were affected both by sentiments about the actor's demeanor and sentiments about the action out of context. Considerably more variance could be explained by entering these elements separately than by regressing on the EPA profile predicted for the combination of demeanor and action, implying that demeanor and action do not form an amalgam that operates in a unitary way within impression-formation processes.

Rashotte (2003) conducted another study of 64 events in a Graeco–Latin square design, this time with stimuli presented to respondents in two different forms: written and videotaped encounters of actors. Separate sets of respondents were employed to rate each kind of stimulus. Some of the events included in this study were:

The lady wrinkles her nose and greets the warden.
The buddy punches and mocks the wrongdoer.
The assailant speaks in a quavering voice and flatters the lady.
The alcoholic blinks and contemplates the flirt.
The clerk leans back and appeases the aunt.

In-context impressions created by videotaped events were substantially more variable than those generated by text descriptions of the events. Evidently, respondents construed what they were seeing in different ways, since the videoed activities were not interpreted for them verbally. Such individual differences constitute error in a study of norms (as explained in the second half of this book), and they reduced the reliability of in-context event elements in videoed events.

Rashotte found that almost all of the impression-formation processes that she examined were more predictable from the text-based data than from the video-based data. Of course, that naturally would follow from the differences in reliability of the in-context ratings based on text and video representations of events. Rashotte correlated regression coefficients in impression-formation equations obtained with text stimuli and regression coefficients obtained with video stimuli. From these correlations she concluded (2003, p. 292) that the relative contributions of event components were very similar for the two stimuli types in predicting evaluations, and moderately similar in predicting potencies and activities of outcomes (except predictions of behavior activities had almost identical patterns for visual and written stimuli). Thus, impression-formation processes were largely parallel in textual and videoed presentations of events.

4.2.5 Settings

Lynn Smith-Lovin (1979, 1987a) initiated impression-formation studies regarding the impacts of settings on impression formation, and the impact of events on impressions of settings. Her 1987 report was based partly on data from the UNC project (described above), and also on a study with undergraduate respondents from the University of South Carolina. The South Carolina study used the same scales as were employed in the UNC project, and it presented the same stimuli for rating, except that some stimuli were phrased in present tense rather than in past tense (e.g., "a flirt flatters a snob" instead of "a flirt flattered a snob").[4]

Smith-Lovin (1978, pp. 83, 86) described the design of her study as follows.

> All possible combinations of EPA values for Actor, Behavior, Object and Setting could not be represented because of the large number of event descriptions which would be generated by a completely orthogonal design (2816 events). The evaluation and activity levels were made orthogonal because this is the area where most interactions have occurred in past studies.
>
> Every possible combination of both positively and negatively evaluated Actors, Behaviors, Object-Persons and Settings and of both active and inactive Actors,

[4]Smith-Lovin found a significant effect associated with tense, but nevertheless, a model ignoring tense fit the data very well, so she dropped consideration of tense in the interests of parsimony (Smith-Lovin 1987a, note 6).

> Behaviors, and Settings appears in one event description. . . . This produced the 128 event descriptions. The activity level of the Object-Person and the potency of all event elements (A, B, O and S) were assigned randomly within each cell of the design.
>
> Two neutral event descriptions were combined with non-neutral settings to produce a total of 130 events.

Vignettes were constructed with two main phrases, and the element to be rated was underlined. For instance, one vignette was "A heroine and a child were together in a riot, and the heroine rescued the child." In this presentation, respondents were to rate their impressions of the heroine in the event. Other respondents rated their in-context impressions of the riot, the child, and the action of rescuing.

Smith-Lovin (1987a, pp. 87–90) found that explicit consideration of the setting of an event caused the setting to enter into impression formation in a number of important ways. Good, lively settings enhanced the goodness of interactants, and active settings enhanced the activity of an actor. Moreover, actors were devaluated if their activity was inconsistent with the tempo of the setting (e.g., an introvert at a party). Behaviors seemed nicer when they occurred in lively low-power scenes such as a party or a playground; behaviors seemed more potent in bad, noisy places such as a mob or riot; and behaviors seemed more lively in lively places. Object persons lost less power and seemed livelier in pleasant settings.

Smith-Lovin (1987a, p. 92) also found that events affected impressions of the settings in which the events occur. Specifically, the significant effects that she found all involved the pre-event feelings about the setting, or else pre-event feelings about the behavior, or else interactions between pre-event feelings about the behavior and pre-event feelings about the object person.

Herman Smith (2002) replicated Smith-Lovin's basic design in Japan, creating two Japanese equivalents for each of Smith-Lovin's 130 sentences, for a total of 260 vignettes describing events situated in specific settings. The actors, behaviors, object persons, and settings in these events were rated in context on Evaluation, Potency, and Activity scales presented with the Japanese version of the *Attitude* data-gathering program (see Chapter 2 for details on the scales and the program). Respondents were 500 students, equally male and female, from several universities in the greater Tokyo area.

Smith (2002) found substantial differences between Japanese and American impression-formation processes when settings are specified. For one thing, consequential gender differences appeared in Japanese equation estimations, whereas gender differences in American equations were minor. Additionally, Japanese equations for predicting outcome feelings about the actor, behavior, and object of an event contained many fewer interaction terms than the corresponding American equations. Smith (2002) concluded from this that "psychological consistency is less important to Japanese than Americans." On the other hand, in Japan as compared to the United States, more interaction terms contributed to setting evaluations, indicating a more nuanced consideration of settings in Japan. Overall, the results suggested that Japanese impression formation relies more on situational factors than on actors and object persons, which accorded with authoritative treatises on East–West differences.

4.3 ATTRIBUTE–IDENTITY AMALGAMATIONS

Harry Gollob's studies concerning impression formation from events incorporated modifiers into the vignettes describing events. For example, his first study (Gollob 1968) used vignettes of the form, "The kind man praises communists," and the Gollob and Rossman study (1973) used vignettes of the form, "Bill helped the corrupt senator." Gollob assumed that a modifier–noun combination creates a psychological amalgam with a unitary sentiment, and the success of his experiments on impression-formation processes suggested that this is true.

Averett (1981; Averett and Heise 1987) undertook a project to examine how such sentiment amalgams are formed. Averett hypothesized that the unitary sentiment associated with an amalgam is generated from sentiments associated with the two components of the amalgam: the modifier and the identity. Averett argued that this impression-formation process, like impression formation from events, should be specifiable with equations describing how the component sentiments sum and interact in producing an outcome.

Averett examined the amalgamation process empirically with a study designed as follows (Averett 1981, p. 39):

> Stimuli consist of 192 modifier–identity combinations representing three replications of a complete 8 × 8 design involving all positive and negative combinations of evaluation, potency, and activity for modifiers and identities. . . . Three stimulus replications of the complete 8 × 8 design were created. Two replications involve stable personality characteristics or traits while the other contains temporary mood states. . . . Data from another study conducted by the author and Lynn Smith-Lovin were available. This dataset contains attribute–identity combinations linking six social characteristics (young, old, male, female, rich, and poor) with [90] occupations. . . . Stimulus frames for semantic differential ratings were "a(n) adjective noun," e.g. "a friendly boyfriend" or "a belligerent outlaw."

Overall, Averett's study presented 275 attribute–identity stimuli to respondents: 121 trait–identity combinations, 90 status–identity combinations, and 64 emotion–identity combinations. The study used the scales, respondents, and questionnaires of the UNC project, as described previously.

Averett and Heise (1987, pp. 107–112) reported that about three-quarters of the variance in mean EPA ratings of modifier–noun combinations could be predicted from the sentiments associated with the modifier and with the noun. For example, the Evaluation of an attribute–identity amalgam was largely predictable by averaging the Evaluation of the attribute and the Evaluation of the identity, with the attribute weighted more. An interaction effect also contributed such that evaluatively consistent combinations such as a virtuous boyfriend and an embarrassed blabbermouth seemed nicer than the average of the components. Weighted averaging also worked for Potency and Activity outcomes, and while some EPA interactions additionally influenced the Potency of an amalgam, no such interactions were evident for Activity. Equations for predicting sentiments associated with emotion–identity combinations were somewhat different than equations for trait–identity combinations; and equations for occupational identities were somewhat different than equations for other kinds of identities. Equations were also somewhat different for male and female respondents.

Averett additionally conducted a study in which event vignettes were created using attribute–identity combinations, in the manner of Gollob.

> Two sets of 64 sentences were created which describe an event in which either the actor or the object was a modified identity. . . . Each of the sets represents an 8 × 8 Graeco–Latin square in which modifier EPA profiles and identity EPA profiles are orthogonal (every modifier profile appears with every identity profile) while either actor and behavior or behavior and object profiles appear once in every row and column. The assignment of profiles to a particular position in the matrix is randomized within the constraints of the Graeco–Latin square design. . . . In addition, two neutral sentences were included to make totals of 65 MABO and 65 ABMO event descriptions.... In both the MABO and ABMO sentence stimuli the in-context evaluation, potency, and activity of the actor was rated. Examples of stimulus frames are "The sentimental daughter oppressed the son" and "The judge contemplated the charming gambler," with the underlined word being rated in each sentence type. (Averett 1981, pp. 39, 41)

Analyses of the data from these events with complexly defined interactants indicated that amalgams act in impression formation in essentially the same way as unitary identities. "The overall structure of event dynamics is similar whether identities are modified or unmodified, and amalgamation processes have a similar structure whether occurring in isolation or in . . . an event description. A composite impression of actor or object acts essentially like a simple identity in reactions to an event" (Averett and Heise 1987, p. 118).

Neil MacKinnon replicated the Averett study in Canada. In a terse report of his equation estimations he concluded that "while one or more coefficients in each equation may be a little smaller or larger than those in the other study, the Canadian amalgamation equations are structurally similar to the American" (MacKinnon 1988, p. 4).

Lisa Thomas and I conducted an amalgamation study that substantially increased the number of stimuli used in estimating equations for predicting EPA values for emotion–identity amalgams. "Only a partial factorial design was possible [because] English provides few emotion labels for states of weak goodness and no words at all for states of activated weak goodness. Thus we could define only seven EPA variants of emotions. Combining seven variants of emotion with eight EPA variants of identity yields 56 possible EPA configurations; replicating four times produces our total of 224 emotion-identity combinations" (Heise and Thomas 1989, p. 142). The Heise–Thomas study used the *Attitude* program for data gathering (see Chapter 2), with approximately 40 female undergraduates and 40 male undergraduates serving as respondents for each stimulus. Analyses were conducted with the emotion–identity data alone, and also with a pooled dataset combining the new Heise–Thomas data with Averett's dataset.

Results obtained in the Heise and Thomas study confirmed the essential findings of Averett's work, with high prediction accuracy (Heise and Thomas 1989, p. 147): "Predicted values from Equations (5–7) correlate highly with the mean empirical ratings of our 224 emotion–identity combinations: 0.94, 0.91, and 0.95 for evaluation, potency, and activity respectively (means were computed across both males and females)." Moreover, some complexities that Averett found dissipated when analyses were conducted with the pooled dataset. In particular, Heise and Thomas found that a single set of prediction equations provided adequate predictions for both emotion–identity and trait–identity combinations; and Heise and Thomas found no significant gender differences.

Smith, Matsuno, and Ike (2001) conducted a Japanese replication of the Averett (1981) and Heise and Thomas (1989) studies. They wrote of generating stimuli by crossing the eight EPA patterns for identities with the eight patterns for traits, the seven EPA patterns for emotions,[5] and the six status characteristics of male, female, rich, poor, young, and old—a design with 168 cells. However, their statement (2001) that "data for male and for female raters are pooled (two sets of 88 stimuli × rater's sex) for a total of 352 observations" suggests that they sampled from the cells of the full factorial design. Their respondents were 30 male and 30 female university students in the greater Tokyo area. Ratings were collected with the Japanese version of the *Attitude* program, whose scales and features are discussed in Chapter 2.

Smith, Matsuno, and Ike (2001) concluded that for Japanese, personal modifiers are more important than identities in forming impressions, that Japanese males and females process modifier–identity combinations differently, and that modifiers and identities interact in more ways during impression formation for Japanese than for Americans.

Tobias Schröder's 2007 study in Germany using the *Surveyor* program included a component devoted to estimating German amalgamation equations.[6] His design crossed the eight EPA configurations for identities with the eight for traits and with eight for emotions, yielding 128 combinations. One stimulus was created for each configuration: for example, *ein bescheidener*

[5]Japanese, like English, lacks an emotion term for the pleasant, activated, and vulnerable states.
[6]See Chapter 3 for details on the German version of the *Surveyor* program that Schröder used and for more information on his respondents. My discussion above of Schröder's amalgamation study is based on an incomplete manuscript that Schröder kindly provided in 2009.

Fischer (a modest fisherman—trait E+P–A– with identity E+P+A–), and *eine verärgerte Dame* (an angry lady—emotion E–P+A+ with identity E+P–A–).

The structure of the German results paralleled those obtained in the United States, although the coefficients had different values. The equations predicting feelings about the amalgam from sentiments associated with attribute and identity accounted for 78 percent or more of the variance in amalgam ratings. No significant differences between male and female prediction equations were found.

4.4 EVENT LIKELIHOODS

A number of traditions of psychological research focus on respondents' likelihood assessments of situations presented in vignettes. For example, Daniel Kahneman and his colleagues had respondents judge the likelihoods of events described in vignettes in order to evaluate propositions about how a sense of the normative emerges in individual minds (e.g., see Kahneman and Miller 1986; Kahneman, Slovic, and Tversky 1982). Harry Gollob and his colleagues (Gollob 1974a, 1974b; Gollob and Fischer 1973; Gollob, Rossman, and Abelson 1973) examined how implicit biases regarding happenings influence respondents' likelihood judgments about someone's behaviors or attributes. These traditions of psychological research were concerned with establishing psychological principles rather than with examining cultural regularities, so they are not directly relevant here. However, Gollob's work requires some detailed consideration since it influenced studies that focused on cultural regularities.

Gollob argued that a simple sentence describing a happening consists of a subject (S), verb (V), and object (O), and each component may be either positive or negative. "Whenever S, V, or O is positive, or the product SV, SO, VO, or SVO is positive, the specified sentence type is said to possess the bias in question. For example, sentence type + + – has S bias, V bias, and SV bias and does not have any of the other biases" (Gollob 1974b, p. 287).

Gollob proposed that biases in a sentence affect the judged likeliness of particular kinds of inferences. For example, start with the core sentence "The kind man hates Mr. B who is a psychologist"—presumably a + – + sentence type in terms of evaluations. A respondent might be given prior information that the man hates Mr. B who is a psychologist, and asked how likely is it that the man is kind.[7] Judging the man as kind completes an S bias and an SO bias, so that is a more likely response than judging the man as unkind, which forms a sentence with no positive biases beyond the O bias in the original information. In the case of likelihood judgments regarding behavior, the identities of

[7]Gollob (1974a) had his respondents make their judgments of likelihood on a graphic rating scale ranging from "extremely improbable" to "extremely probable," with numbers from –7 through zero to +7 arranged equidistant along the scale.

the actor and object were established first, then respondents were asked to rate the likelihood of a specific behavior: for example, "Bill is unfriendly and Joe is friendly. How probable is it that Bill likes Joe?" Predictions regarding the behavior's likelihood derive from the V, SV, VO, and SVO biases, since these are the relevant ones for verb inferences.

The various biases are weighted differentially in predicting an inference. Any bias that does not relate directly to the element being inferred is weighted zero, as seen above. Specific weights for the relevant biases must be estimated empirically, and the empirical weights vary depending on the sentences used in a particular study. To illustrate, one weight would be appropriate for predicting inferences with a psychologist as a positive object and a bill collector as a negative object, and another weight probably would be required with a psychologist as a positive object and a burglar as a negative object. Weights also would change if the objects were not persons but issues, such as opinions about sweatshops versus cruises. Weights additionally might vary across individual observers.

> Social inference results are complex and ... the particular types of content and types of inferences being made affect which biases are important and the interpretation of those biases. Fortunately, however, over a fairly wide range of item sets, usually only two or three biases have large effects on the inferences made for any one item set. The complexity enters the picture when one attempts to determine *a priori* just which biases are important to perceivers and what interpretations of the biases are appropriate for the content and type of judgment involved. (Gollob 1974b, p. 320)

Gollob and Fischer (1973) proposed a somewhat different hypothesis regarding behavior inferences: that subjective likelihoods depend on amounts of change in impressions. Relevant impressions were generated in a syllogistic-like structure: A given behavior, *g*, was offered as a true premise which produced one impression, and the respondent then rated the likelihood of the actor engaging in an independent behavior, *i*, that generated a different impression. "The greater the change that ... Act *i* would require one to make in his impression of a person based on the given statement that an actor engages in Act *g*, the lower the likelihood that one will judge the statement to be inferred as true" (Gollob and Fischer 1973, p. 16). Gollob and Fischer examined impressions on the Evaluation, Potency, and Activity dimensions and found that changes in evaluative impressions were notably related to likelihood judgments, but changes in impressions on the other two dimensions had much less impact.

Neil MacKinnon and I (Heise and MacKinnon 1987) treated subjective likelihoods of events as culturally normative outcomes of affective processes. The general idea was that an event description recruits a set of cultural sentiments; the event's structure mixes these sentiments to create impressions diverging to some degree from the original sentiments; and the greater the divergence, the less likely the event seems. "Event likelihoods arise in affect control theory because a key premise of the theory is that events are constructed so as to minimize the deflections of transient feelings from fundamental sentiments, and a behavior that so minimizes deflections is the expected or intended behavior—the likely behavior—in the circumstances" (Heise and MacKinnon 1987, p. 135).

The Heise and MacKinnon study examined the 515 event sentences used in the UNC impression-formation study described above. Likelihood ratings for each event had been obtained on a seven-position scale with the positions labeled zero to six, and endpoints defined by the phrases "not at all likely" and "extremely likely." Ratings were coded with an interval-level metric derived through successive-intervals scaling, and mean likelihoods were computed for each event across male and female raters separately, with 19 or more of each gender.

The first task was to examine whether deflections were inversely related to likelihood, bigger deflections from an event implying lower rated likelihood.

> Deflections are estimated from the same measures as were used to develop equations describing affective dynamics. Evaluation, Potency, and Activity (EPA) measures were obtained for the Actor, Behavior, and Object (ABO) of each event. One set of ratings describes the ABO elements when presented out of the context of any sentence and another set of ratings describes them when presented together in the context of a given event sentence. Differences between out-of-context and in-context ratings assess how much the event deflects feelings away from fundamental sentiments. (Heise and MacKinnon 1987, p. 135)

Plotting event likelihoods against deflections revealed that the relation was severely heteroscedastic, with events of many likelihoods occurring at low deflection levels. Nevertheless, the plots provided support for the hypothesized deflection–likelihood relation. Events that generated extremely large affective deflections all were viewed as improbable and only events producing small affective deflections were seen as extremely likely.

Correlational methods were used to examine the relations between likelihood and the nine deflections in a simple event description: the deflection for actor evaluation, the deflection for actor potency, the deflection for actor activity, the deflection for behavior evaluation, the deflection for behavior potency, the deflection for behavior activity, the deflection for object evaluation, the deflection for object potency, and the deflection for object activity.

> The affective deflections produced by events generally correlate negatively with event likelihoods, indicating that events which produce greater amounts of affective change have lower likelihoods, as is expected theoretically. . . . The magnitudes of the correlations are not great, but this is at least partly because correlation coefficients are inappropriate for the kind of relation that is involved. Additionally, magnitudes of some of the correlations are limited by the restricted ranges of

> the deflections. Multiple regression analyses revealed that more variance in likelihood ratings is explained by the set of nine affective deflections than by any one of them alone. Thus the impacts of deflections cumulate, as is required in affect control theory. (Heise and MacKinnon 1987, p. 148)

Heise and MacKinnon found that deflections accounted for about one-third of the variance in event likelihoods. The figure was not higher because many events that yielded low deflections nevertheless were rated as improbable. A search for ways to improve the predictability of event likelihoods yielded a number of interesting results.

As discussed above, Gollob's biases performed well as predictors, accounting for just under half of the variance in event likelihoods. All seven biases were significant predictors of likelihood in analyses of male data, and all but the S and V biases were significant in the female data. Significant S, O, and SO biases imply that respondents were basing their likelihood judgments on the kinds of actors and object appearing in events, not simply on the likelihood of a given behavior for predefined interactants. For example, the event "The schoolgirl tickled the gambler" was rated as much less likely than predicted by deflections, and in this case the gambler identity gave a negative value to the O and SO biases, meaning that respondents presumably thought an event involving a gambler was unlikely, and especially so when the other interactant was a schoolgirl.

Institutionalization moderated the relation between deflections and likelihood.

> When an actor's identity is institutionally vague, deflections predict likelihoods but the level of predictability is low. Predictability increases to moderate levels for events involving actors who are deviant with respect to standard social institutions. High levels of predictability are attained for events that involve people acting in roles that are central and normal in standard institutional contexts. (Heise and MacKinnon 1987, p. 149)

Institutionally clear events involved actors in conventional roles within the family, legal, medical, and other institutions. Deviant actors were those whose institutional identities had negative evaluations. Institutionally vague actors included identities such as child, lover, and hero; blabbermouth, smart-aleck, and loafer. Within the set of 180 institutionally clear events, deflections accounted for 61 percent of the variance in males' likelihood judgments and 44 percent of the variance in females' likelihood judgments.[8]

MacKinnon's Canadian study (1985) also examined the relation between deflections and rated likelihoods. However, the behaviors in his study were frequency modified: (e.g., "The physician occasionally avoids the intern"), so

[9]In retrospect, it seems evident that the institutionalization moderator might be interpretable in terms of Gollob's biases, but the Heise and MacKinnon (1987) study did not address this connection.

they relate more to information summaries than to observed events. However, he did find the expected inverse relation between deflections and rated likelihood, with a cumulation of deflection effects, for each relation between deflections and likelihood that was significant (MacKinnon 1985, Table 11).

Several different models for predicting likelihood ratings of the 515 events in the UNC study were considered in my examination of affect control theory's formal model (Heise 1985), a study conducted after the Heise and MacKinnon (1987) work. In all models, coefficient estimates were constrained to equality across genders. A regression on the nine deflections separately explained 34 percent of the likelihood variance; a regression with equality constraints for the three actor deflections, the three behavior deflections, and the three object deflections explained 29 percent of the likelihood variance; and a regression with all nine coefficients constrained to equality explained 27 percent of the variance. A fourth model was based on the fact that a deflection is a squared difference (transient impression minus fundamental sentiment, squared) and thus can be disaggregated into three terms: the transient squared, the fundamental squared, and the cross-product of transient with fundamental. Likelihoods were regressed on the terms from the disaggregation, with coefficients for the nine different deflections constrained to equality. My report on this model, which accounted for 43 percent of the likelihood variance, was as follows (Heise 1985, p. 208): "The signs are as expected for a disaggregation. The coefficient for the cross-product term is about twice the coefficient for the squared fundamental, as required in a squared difference. However, the coefficient for the squared transient is about half the value it would have in a squared difference, and this deviation allows the proportion of explained variance to jump sixteen points above the *R*-square for equally-weighted deflections without disaggregation."

Another model considered in the Heise (1985) study combined deflections with the Gollob biases relating to evaluations. This model explained 53 percent of the variance in likelihoods. Deflections (with their regression coefficients constrained to equality) were significant predictors. So were the biases relating to actor and object (S, O, and SO), and biases relating to consistency of behavior with actor and object (SV, VO, and SVO). The behavior bias (V) was significant, but its sign was negative, indicating that respondents thought good behaviors were less likely than bad behaviors.

I tried to incorporate the Gollob biases into the affect control theory model for predicting social interaction. This resulted in some plausible simulation results, but mostly created severe problems.

> [The combined deflection and bias model] frequently generates EPA profiles that are outside the range of measured profiles for behaviors. When behaviors are retrievable, they are not always plausible. Additionally [this model] causes simulations to be unstable. Behaviors are selected that cause person transients to become more extreme. That, in turn, causes later behavior profiles to be more extreme, until after a number of cycles of interaction, the generated behavior profiles always are beyond the empirical range of dictionary behaviors. (Heise 1985, p. 211)

After trying simulations with all of the models that I had estimated, I concluded that equally weighted deflections performed best in parameterizing the simulation system, even though the equally weighted deflection model accounted for the least variance in the empirical predictions of likelihood.

> This implementation attaches more import to successful simulations than to statistics. If that seems seditious in quantitative sociology, it is only because we so rarely have the choice. Do you parameterize your model with statistics that best fit a given data set but which make simulations impossible, or use numbers that have a weaker statistical basis but which allow you to simulate many phenomena of interest? Simulations win. No one wants a model that fits data from one study but applies to nothing else. (Heise 1985, p. 221)

My central thought regarding why the addition of biases to deflections wrecks successful simulations was that judging the likelihood of an event and judging the plausibility of a simulation outcome actually involve different dimensions.

> Perhaps the relevant dimension in simulations is not Likely versus Unlikely but Appropriate–Natural–Expected versus Inappropriate–Irregular–Unexpected. . . . Following this line of thought, deflections would be the predominant predictors of appropriateness, but deflections would predict likelihood measurements only to the extent that likelihood and appropriateness correlate. (Heise 1985, p. 220)

Kahneman and Miller (1986, p. 137) suggested that a sense of normality reflects a lack of surprise rather than an assessment of high probability. "Probability is always construed as an aspect of anticipation, whereas surprise is the outcome of what we shall call backward processing: evaluation after the fact. Probability reflects expectations. Surprise (or its absence) reflects the failure or success of an attempt to make sense of an experience."

These considerations suggest that deflections produced by events may predict ratings on a scale of normal to surprising better than they do on a scale of likelihood, and the same presumably would be true in judging events of social interaction, either real or simulated.

4.5 SYNOPSIS

Vignette studies get at unconscious cultural processing. In the factorial-surveys tradition, vignettes have been used to develop explanations of how people assign social status, judge the seriousness of someone's deviance, weigh the fairness of wages, and determine obligations owed to kin. In the impression-formation tradition, vignettes have been used to derive models of how people integrate feelings in actor–behavior–object events, actor–

behavior–object–setting events, self-directed behavior, and attribute–identity combinations.

Vignettes are composed by crossing variables believed to affect outcomes, and the ideal is a full factorial design with at least one vignette in each cell of the typology created by the crossed variables. However, full factorial designs explode rapidly in size as more compositional variables are considered and as variables are divided into more levels. Two methods have been employed to reduce full factorial designs so as to keep vignette studies economically feasible. Factorial surveys often draw random samples of vignettes from the full factorial design, an approach that is appropriate when it reasonably can be assumed that only main effects of compositional variables contribute to outcomes, not interactions among the compositional variables. Impression-formation studies often condense factorial designs systematically, via Latin squares or Graeco–Latin squares, thereby permitting investigation of main effects and some interactions among variables, although not all interactions.

Vignette sampling has been applied in impression-formation studies when a full factorial design is not economically feasible, but since random sampling may preclude finding interesting interactions, a systematically condensed set of vignettes is far preferable. In the appendix to this chapter I present a double Latin square design that may be used in future impression-formation studies of actor–behavior–object events. The appendix also provides some sample designs for studying attribute–identity combinations.

Discovery of intercultural differences is one reason to conduct studies of unconscious cultural processing. Research in the factorial-surveys tradition has dealt with this issue mainly via analyses intended to reveal variant norms among persons differing on characteristics such as gender, race, family status, and family size. Researchers in this tradition have discovered remarkably few subcultures associated with such demographic distinctions. Impression-formation research has identified some subcultural variations with regard to gender, mainly in the relatively gender-segregated society of Japan. On the other hand, because impression-formation research has a database of past results, new results are routinely compared and contrasted with outcomes attained in other societies and at other times, and some notable differences in cultural processing have been identified.

Like studies meant to build repositories of sentiments, vignette studies present each respondent with only a portion of the total number of stimuli being considered in a study. Two methods have been adopted to build sets of vignettes for each respondent. One possibility is to present each respondent with a separate random sample from the total set of vignettes; Rossi and Rossi (1990) and others in the factorial-surveys tradition have used this approach. A second approach is to assign respondents randomly to one of several subsets of vignettes "chosen systematically ... in such a fashion as to insure the statistical equivalence of the samples", as described by Berk and Rossi (1977, p. 128), who used the method. Studies of impression formation have used the second method.

4.6 CHAPTER HIGHLIGHTS

- Implicit cultural processes can be uncovered by presenting culturally interpretable situations in vignettes, asking respondents to rate each case along some dimension and then analyzing the data to reveal which features of the vignettes were influencing respondents' ratings.
- One tradition of vignette studies, factorial surveys, has explored how situational features influence normative judgments. Each study begins by combining every level of relevant features in order to generate a hypothetical universe of vignettes. Then questionnaires for respondents are obtained by sampling from the universe. Ratings of respondents are pooled to yield a large sample of vignettes for analyses.
- Data from factorial surveys have been analyzed in several different modes, all of which employ the assumption that features do not interact extensively in generating responses to objects.
- An impression-formation study presents respondents with brief vignettes describing events in which various kinds of actors, behaviors, object persons, and sometimes settings have been combined. Respondents rate each constituent of the event on Evaluation, Potency, and Activity (EPA) scales. The in-context ratings are regressed on out-of-context EPA ratings of the constituents in order to discover how respondents fuse their pre-event feelings in arriving at post-event feelings.
- A large impression-formation study was conducted in the United States in the 1970s. Equations for predicting impression formation from events were estimated from ratings of 515 event vignettes. The study indicated that impression formation processes are complex, with nine first-order terms, 14 two-variable interactions, and five three-variable interactions appearing in some of the equations.
- Additional impression-formation studies have been conducted in Canada, Japan, and Germany.
- Impressions from attribute–identity amalgamations involve confronting respondents with a stimulus in which a person in some identity has some kind of attribute. EPA ratings of the combination of identity and attribute are obtained and analyzed in order to determine how the combination produces a new feeling about the person.
- Studies of attribute–identity amalgamations have been conducted in the United States, Canada, Germany, and Japan.
- Researchers in several disciplines have employed vignettes to examine the problem of predicting respondents' judgments of event likelihoods. Psychological approaches do well in predicting subjective likelihood. A sociological model seemed more adapted to predicting judgments of surprising versus normal.

4.7 APPENDIX: IMPRESSION-FORMATION STUDY DESIGNS

4.7.1 Impressions from Events

The appearance of interaction terms such as B_eO_e and B_eO_p in early impression-formation research suggested that an adequately designed exploratory study of impression-formation processes should include sets of events to manifest every conceivable interaction of different event components, both within EPA dimensions and across dimensions. The size of a fully crossed set is 8·8·8 = 512 events. Such a study was conducted once in the United States, but in general a stimuli set of 512 events constitutes a study that is excessively costly. The actor, behavior, and object person in each event must be rated on the evaluation, potency, and activity dimensions, so respondents must perform 4,608 ratings of event constituents (nine ratings per event times 512 events). Additionally, respondents must provide out-of-context ratings for the identities and behaviors used to construct event descriptions. Thus, about 5,000 ratings must be acquired. Each rating must be replicated about 25 times to deal with measurement unreliability, so a total of 125,000 ratings must be collected from respondents. Figuring conservatively that respondents can do 300 ratings in an hour, more than 400 hours of respondent time are required, and this commitment must be obtained from each group being studied (e.g., from females and then again from males).

The number of ratings required can be reduced substantially by adopting the affect control theory assumption that different types of actors and different types of behaviors have no unique effects on impression formation, where types are defined by combinations of two or more EPA dimensions. For example, impression formation dynamics are assumed to be the same for good and potent actors as for bad and weak actors, after allowing for the general effects of goodness versus badness and of potency versus impotency. The biggest regression equation to be considered would therefore consist of a constant, nine first-order terms (e.g., A_e, A_p, A_a, B_e), 27 second-order terms (e.g., A_eB_e, A_pB_e), and 27 third-order terms (e.g., $A_eB_eO_e$, $A_eB_aO_p$). That gives 64 coefficients to be estimated, and a minimal study must consider 64 events in order to estimate every coefficient, although a greater number of events permits more dependable estimations.

To determine which crossings of actor, behavior, and object EPA types are most involved in impression formation, I tallied the interactions found in nine regression estimations of impression-formation equations (male and female estimations in the United States, Canada, Japan, and China, and a combined-gender analysis in Germany).

- Forty-three interactions involved Evaluation alone, or Potency alone, or Evaluation crossed with Potency.
- Thirty-one interactions involved Potency alone, or Activity alone, or Potency crossed with Activity.

- Twenty interactions involved Evaluation alone, or Activity alone, or Evaluation crossed with Activity.

These results suggest that valuable information might be gained by fully crossing all ABO combinations of Evaluation and Potency, and by fully crossing all ABO combinations of Potency and Activity. Doing so would produce two sets of 64 sentences, the total of 128 being double the minimum requirement of 64. Crossing two dimensions leaves the third dimension undefined, but the third dimension can be added as a Latin square. In particular, ABO combinations of Activity can be imposed as a Latin square on top of the crossing of Evaluation and Potency, and ABO combinations of Evaluation can be imposed as another Latin square on top of the crossing of Potency and Activity.

The ABO combinations of Evaluation are:

(+ _ _, + _ _, + _ _)

(+ _ _, + _ _, − _ _)

(+ _ _, − _ _, + _ _)

(+ _ _, − _ _, − _ _)

(− _ _, + _ _, + _ _)

(− _ _, + _ _, − _ _)

(− _ _, − _ _, + _ _)

(− _ _, − _ _, − _ _)

Each group gives the Evaluation pattern for actor, behavior, and object. Potency and Activity are unspecified and represented by underscores (_), since this definition concerns the patterns for Evaluation alone.

The ABO combinations of Potency are:

(_ + _, _ + _, _ + _)

(_ + _, _ + _, _ − _)

(_ + _, _ − _, _ + _)

(_ + _, _ − _, _ − _)

(_ − _, _ + _, _ + _)

(_ − _, _ + _, _ − _)

(_ − _, _ − _, _ + _)

(_ − _, _ − _, _ − _).

In this case, Evaluation and Activity are unspecified and represented by underscores (_).

The ABO combinations of Activity are:

(_ _ +, _ _ +, _ _ +)

(_ _ +, _ _ +, _ _ −)

(_ _ +, _ _ −, _ _ +)

(_ _ +, _ _ −, _ _ −)

(_ _ −, _ _ +, _ _ +)

(_ _ −, _ _ +, _ _ −)

(_ _ −, _ _ −, _ _ +)

(_ _ −, _ _ −, _ _ −).

Here Evaluation and Potency are unspecified and represented by underscores (_).

Now it is possible to define the first set of 64 events, in which Evaluation ABOs are crossed with Potency ABOs, and Activity ABOs are imposed as a Latin square. [The Latin square used here was obtained from an algorithm provided by Bogomolny (2008).] The event configurations are given in the top half of Table 4.2, with an example sentence for each configuration, based on sentiments measured in the United States.

The second set of 64 events crosses Potency ABOs with Activity ABOs, and Evaluation ABOs are imposed as a Latin square. The same Latin square is applied here, but it has been reversed, with its patterns going from bottom to top rather than top to bottom, in order to eliminate duplicate patterns in the overall stimulus set. The second half of Table 4.2 shows the resulting event configurations and gives an example sentence for each configuration.

The 79 identities and behaviors used to construct the event sentences in Table 4.2 are given in Table 4.3. Some illustrative frames for presenting in-context event stimuli and out-of-context identities and behaviors are displayed in Table 4.4.

The set of events defined in Table 4.2 have excellent characteristics as bases for regression analyses. First-order predictors (e.g., A_e, A_p, A_a, B_e), second-order predictors (e.g., A_eB_e, A_pB_e), and third-order predictors (e.g., $A_eB_eO_e$, $A_eB_aO_p$) all are orthogonal, providing optimal efficiency in regressions. (In practice, predictors will have some small correlations because real-life identities and behaviors do not implement the designs perfectly.) Multiple events in the list represent every relevant configuration for estimating cross-term interactions. For example, estimation of an A_eB_e interaction depends on four configurations of events: (+ _ _, + _ _, _ _ _), (+ _ _, − _ _, _ _ _), (− _ _, + _ _, _ _ _), and (− _ _, − _ _, _ _ _); and the set of events contains

TABLE 4.2 Event Patterns, with Example Sentences

E Crossed with P; A Latinized	
E+P+A+, E+P+A+, E+P+A+	an athlete is entertaining a teammate.
E+P+A+, E+P+A+, E+P–A–	a registered nurse is coaching an applicant.
E+P+A+, E+P–A–, E+P+A+	an athlete is watching a champion.
E+P+A+, E+P–A–, E+P–A–	an athlete is following an old-timer.
E+P–A–, E+P+A+, E+P+A+	a retiree is laughing with an athlete.
E+P–A–, E+P+A+, E+P–A–	an interviewee is surprising an old-timer.
E+P–A–, E+P–A–, E+P+A+	a beginner is obeying a champion.
E+P–A–, E+P–A–, E+P–A–	a retiree is waiting on a patient.
E+P+A+, E+P+A+, E–P+A–	an athlete is competing with an enemy.
E+P+A+, E+P+A+, E–P–A+	an athlete is joking with a hothead.
E+P+A+, E+P–A–, E–P+A–	a heroine is following a murderer.
E+P+A+, E+P–A–, E–P–A+	a registered nurse is watching a brat.
E+P–A–, E+P+A+, E–P+A–	an old-timer is laughing with an executioner.
E+P–A–, E+P+A+, E–P–A+	an innocent is entertaining a psychotic.
E+P–A–, E+P–A–, E–P+A–	a retiree is obeying a serial murderer.
E+P–A–, E+P–A–, E–P–A+	a patient is watching a chatterbox.
E+P+A+, E–P+A–, E+P+A+	an athlete is staring at a teammate.
E+P+A+, E–P+A–, E+P–A–	a registered nurse is monitoring a patient.
E+P+A+, E–P–A+, E+P+A+	an athlete is babbling to a teammate.
E+P+A+, E–P–A+, E+P–A–	a registered nurse is fussing over a patient.
E+P–A–, E–P+A–, E+P+A+	a retiree is monitoring a registered nurse.
E+P–A–, E–P+A–, E+P–A–	an applicant is staring at an interviewee.
E+P–A–, E–P–A+, E+P+A+	a beginner is babbling to a champion.
E+P–A–, E–P–A+, E+P–A–	a retiree is fussing over an old-timer.
E+P+A+, E–P+A–, E–P+A–	a heroine is confining a serial murderer.
E+P+A+, E–P+A–, E–P–A+	a registered nurse is monitoring a brat.
E+P+A+, E–P–A+, E–P+A–	an athlete is interrupting an enemy.
E+P+A+, E–P–A+, E–P–A+	a champion is blabbering to a gossip.
E+P–A–, E–P+A–, E–P+A–	a retiree is monitoring an executioner.
E+P–A–, E–P+A–, E–P–A+	a patient is staring at a psychotic.
E+P–A–, E–P–A+, E–P+A–	an innocent is pinching a murderer.
E+P–A–, E–P–A+, E–P–A+	a retiree is fussing over a brat.
E–P+A–, E+P+A+, E+P+A+	an executioner is laughing with a registered nurse.
E–P+A–, E+P+A+, E+P–A–	a serial murderer is joking with a retiree.
E–P+A–, E+P–A–, E+P+A+	an enemy is following a heroine.
E–P+A–, E+P–A–, E+P–A–	a murderer is waiting on a patient.
E–P–A+, E+P+A+, E+P+A+	a gossip is entertaining an athlete.
E–P–A+, E+P+A+, E+P–A–	a lunatic is coaching an applicant.
E–P–A+, E+P–A–, E+P+A+	a psychotic is following a heroine.
E–P–A+, E+P–A–, E+P–A–	a brat is obeying a an old-timer.
E–P+A–, E+P+A+, E–P+A–	an executioner is joking with a serial murderer.
E–P+A–, E+P+A+, E–P–A+	a murderer is surprising a chatterbox.
E–P+A–, E+P–A–, E–P+A–	a murderer is obeying a serial murderer.

TABLE 4.2 (*Continued*)

E–P+A–, E+P–A–, E–P–A+	a serial murderer is watching a psychotic.
E–P–A+, E+P+A+, E–P+A–	a psychotic is competing with a serial murderer.
E–P–A+, E+P+A+, E–P–A+	a chatterbox is joking with a gossip.
E–P–A+, E+P–A–, E–P+A–	a hothead is obeying a serial murderer.
E–P–A+, E+P–A–, E–P–A+	a hothead is following a lunatic.
E–P+A–, E–P+A–, E+P+A+	an enemy is monitoring an athlete.
E–P+A–, E–P+A–, E+P–A–	a murderer is confining an innocent.
E–P+A–, E–P–A+, E+P+A+	an executioner is fussing over a heroine.
E–P+A–, E–P–A+, E+P–A–	a murderer is interrupting an interviewee.
E–P–A+, E–P+A–, E+P+A+	a psychotic is confining an athlete.
E–P–A+, E–P+A–, E+P–A–	a brat is staring at an old-timer.
E–P–A+, E–P–A+, E+P+A+	a chatterbox is interrupting a teammate.
E–P–A+, E–P–A+, E+P–A–	a lunatic is pinching a patient.
E–P+A–, E–P+A–, E–P+A–	a murderer is monitoring a divorcé.
E–P+A–, E–P+A–, E–P–A+	a divorcé is confining a brat.
E–P+A–, E–P–A+, E–P+A–	a murderer is babbling to a serial murderer.
E–P+A–, E–P–A+, E–P–A+	a divorcé is blabbering to a gossip.
E–P–A+, E–P+A–, E–P+A–	a psychotic is monitoring a divorcé.
E–P–A+, E–P+A–, E–P–A+	a brat is staring at a gossip.
E–P–A+, E–P–A+, E–P+A–	a gossip is interrupting a divorcé.
E–P–A+, E–P–A+, E–P–A+	a psychotic is pinching a hothead.
	P Crossed with A; E Latinized
E–P+A+, E–P+A+, E–P+A+	an outlaw is slugging a bully.
E–P+A+, E–P+A+, E+P+A–	a gangster is chewing out a priest.
E–P+A+, E+P+A–, E–P+A+	a gangster is comforting a gunman.
E–P+A+, E+P+A–, E+P+A–	a gangster is praying with a priest.
E+P+A–, E–P+A+, E–P+A+	a physician is scolding an outlaw.
E+P+A–, E–P+A+, E+P+A–	a genius is chewing out a scientist.
E+P+A–, E+P+A–, E–P+A+	a priest is comforting a gangster.
E+P+A–, E+P+A–, E+P+A–	a physician is massaging a scientist.
E–P+A+, E–P+A+, E+P–A+	a bully is slugging a child.
E–P+A+, E–P+A+, E–P–A–	a villain is slugging a victim.
E–P+A+, E+P+A–, E+P–A+	a lady-killer is snuggling a teenager.
E–P+A+, E+P+A–, E–P–A–	an outlaw is soothing a victim.
E+P+A–, E–P+A+, E+P–A+	a grandparent is scolding a champion.
E+P+A–, E–P+A+, E–P–A–	a grandparent is chewing out a deadbeat dad.
E+P+A–, E+P+A–, E+P–A+	a pastor is snuggling a toddler.
E+P+A–, E+P+A–, E–P–A–	a priest is praying with a dropout.
E–P+A+, E+P–A+, E–P+A+	a gunman is chatting up a gangster.
E–P+A+, E+P–A+, E+P+A–	a villain is querying a physician.
E–P+A+, E–P–A–, E–P+A+	an outlaw is ignoring a gunman.
E–P+A+, E–P–A–, E+P+A–	a gangster is scoffing at a priest.
E+P+A–, E+P–A+, E–P+A+	a grandparent is escaping a gunman.

(Continued)

TABLE 4.2 ***(Continued)***

E+P+A–, E+P–A+, E+P+A–	a physician is chatting up a scientist.
E+P+A–, E–P–A–, E–P+A+	a grandparent is deprecating a bully.
E+P+A–, E–P–A–, E+P+A–	a scientist is abandoning a genius.
E–P+A+, E+P–A+, E+P–A+	a lady-killer is lusting for a child.
E–P+A+, E+P–A+, E–P–A–	a villain is jesting with a loafer.
E–P+A+, E–P–A–, E+P–A+	a bully is ignoring a youngster.
E–P+A+, E–P–A–, E–P–A–	a gangster is deprecating a gunman.
E+P+A–, E+P–A+, E+P–A+	a pastor is chatting up a youngster.
E+P+A–, E+P–A+, E–P–A–	a physician is querying an unemployed person.
E+P+A–, E–P–A–, E+P–A+	a genius is scoffing at a youngster.
E+P+A–, E–P–A–, E–P–A–	a pastor is abandoning an unemployed person.
E+P–A+, E–P+A+, E–P+A+	a teenager is combating a bully.
E+P–A+, E–P+A+, E+P+A–	a youngster is slugging a physician.
E+P–A+, E+P+A–, E–P+A+	a teenager is soothing an outlaw.
E+P–A+, E+P+A–, E+P+A–	a child is snuggling a grandparent.
E–P–A–, E–P+A+, E–P+A+	a do-nothing is scolding a bully.
E–P–A–, E–P+A+, E+P+A–	a deadbeat dad is chewing out a grandparent.
E–P–A–, E+P+A–, E–P+A+	an unemployed person is soothing an outlaw.
E–P–A–, E+P+A–, E+P+A–	a dropout is comforting a genius.
E+P–A+, E–P+A+, E+P–A+	a teenager is scolding a toddler.
E+P–A+, E–P+A+, E–P–A–	a youngster is slugging a do-nothing.
E+P–A+, E+P+A–, E+P–A+	a teenager is massaging a toddler.
E+P–A+, E+P+A–, E–P–A–	a child is comforting a deadbeat dad.
E–P–A–, E–P+A+, E+P–A+	a deadbeat dad is scolding a toddler.
E–P–A–, E–P+A+, E–P–A–	a do-nothing is chewing out an unemployed person.
E–P–A–, E+P+A–, E+P–A+	a do-nothing is soothing a genius.
E–P–A–, E+P+A–, E–P–A–	a loafer is comforting an unemployed person.
E+P–A+, E+P–A+, E–P+A+	a child is escaping a bully.
E+P–A+, E+P–A+, E+P+A–	a teenager is querying a grandfather.
E+P–A+, E–P–A–, E–P+A+	a youngster is ignoring an outlaw.
E+P–A+, E–P–A–, E+P+A–	a child is submitting to a grandfather.
E–P–A–, E+P–A+, E–P+A+	an unemployed person is chatting up a gunman.
E–P–A–, E+P–A+, E+P+A–	a loafer is escaping a priest.
E–P–A–, E–P–A–, E–P+A+	a victim is submitting to an outlaw.
E–P–A–, E–P–A–, E+P+A–	a deadbeat dad is ignoring a physician.
E+P–A+, E+P–A+, E+P–A+	a teenager is escaping a toddler.
E+P–A+, E+P–A+, E–P–A–	a child is jesting with an unemployed person.
E+P–A+, E–P–A–, E+P–A+	a teenager is scoffing at a toddler.
E+P–A+, E–P–A–, E–P–A–	a child is deprecating an unemployed person.
E–P–A–, E+P–A+, E+P–A+	a deadbeat dad is lusting for a teenager.
E–P–A–, E+P–A+, E–P–A–	a do-nothing is jesting with a dropout.
E–P–A–, E–P–A–, E+P–A+	a loafer is ignoring a toddler.
E–P–A–, E–P–A–, E–P–A–	an unemployed person is scoffing at a deadbeat dad.

TABLE 4.3 Identities and Behaviors Used in the Example Sentences

EPA Pattern	Identities
+++	athlete, champion, registered nurse, heroine, teammate
++−	genius, grandparent, pastor, physician, priest, scientist
+−+	child, teenager, toddler, youngster
+−−	applicant, beginner, innocent, interviewee, old-timer, patient, retiree
−++	bully, gangster, gunman, lady-killer, outlaw, villain
−+−	divorcé, enemy, executioner, murderer, murderess, serial murderer
−−+	brat, chatterbox, gossip, hothead, lunatic, psychotic
−−−	deadbeat dad, do-nothing, dropout, loafer, unemployed person, victim
	Behaviors
+++	coaching, competing with, entertaining, joking with, laughing with, surprising
++−	comforting, massaging, praying with, snuggling, soothing
+−+	chatting up, escaping, jesting with, lusting for, querying
+−−	following, obeying, waiting on, watching
−++	chewing out, scolding, slugging
−+−	confining, monitoring, staring at
−−+	babbling to, blabbering to, fussing over, interrupting, pinching
−−−	abandoning, deprecating, ignoring, scoffing at, submitting to

TABLE 4.4 Frames for Presenting Stimuli When Studying Impressions from Events

Rate how the athlete seems when
an ATHLETE is entertaining a teammate.

Rate how the act of staring at seems when
a patient is STARING AT a psychotic.

An athlete
is

Entertaining someone
is

32 events of each of these types. Estimation of an $A_eB_aO_p$ interaction depends on eight configurations of events: (+ _ _, _ _ +, _ + _), (+ _ _, _ _ +, _ − _), (+ _ _, _ _ −, _ + _), (+ _ _, _ _ −, _ − _), (− _ _, _ _ +, _ + _), (− _ _, _ _ +, _ − _), (− _ _, _ _ −, _ + _), and (− _ _, _ _ −, _ − _); and the set contains 16 events of each of these types.

The vignette sentence for each configuration in Table 4.2 must be presented three times to respondents to acquire EPA ratings of the in-context actor, behavior, and object. Each identity and behavior used to construct the sentences must be presented once for an out-of-context rating. Thus, in the case of the example events based on this design, ratings for 463 stimuli must be obtained—about one-tenth as many stimuli as would be required in a full factorial design.

4.7.2 Impressions from Attribute–Identity Combinations

Personal attributes amalgamate with identities, and impressions of the amalgamations differ from feelings about the constituent attributes and identities. Some examples are: enraged mother, modest mother, rich teacher, relaxed teacher. Four kinds of personal attributes have been considered in affect control theory: emotions (e.g., enraged), traits (e.g., modest), statuses (e.g., rich), and conditions (e.g., relaxed). Most impression-formation research focuses on emotions and traits.

Feelings about personal attributes, identities, and combinations of the two are assessed on the Evaluation, Potency, and Activity (EPA) dimensions. Impressions of attribute–identity amalgamations then are predicted with regression equations that combine out-of-context EPA measures of the attribute and identity and interactions between attribute and identity on the EPA dimensions. The regression equations consist of up to 16 terms: three EPA variables for the attribute, three for the identity, nine cross-term interaction effects, and a constant. Thus, a study of emotion impressions alone or of trait impressions alone requires EPA ratings of at least 16 modifier–identity combinations, and a study of both emotions and traits requires EPA ratings of at least 32 modifier–identity combinations. Substantially more cases are required for reliable parameter estimates.

Personal attributes vary in eight major ways along the EPA dimensions. Table 4.5 lists the patterns and gives examples of a corresponding emotion and of a corresponding trait for each pattern.

Crossing all EPA patterns for attributes with all EPA patterns of identities generates a set of 64 combinations to serve as stimuli in an impression-formation study. Sixty-four emotion–identity combinations could be constructed to study impressions created by those kinds of amalgamations, and 64 trait–identity combinations could be constructed to study impressions of those kinds of amalgamations. Alternatively, emotions and traits both could be included in a set of 64 stimuli by distributing each in a checkerboard pattern, as shown in Table 4.6. The checkerboard design provides data to test

TABLE 4.5 EPA Patterns for Attributes, with Example Emotions and Traits

Pattern	Emotion	Trait
Good, potent, active (E+P+A+)	gleeful	courageous
Good, potent, passive (E+P+A–)	calm	modest
Good, impotent, active (E+P–A+)[a]	emotional	feminine
Good, impotent, passive (E+P–A–)	nostalgic	cautious
Bad, potent, active (E–P+A+)	enraged	ruthless
Bad, potent, passive (E–P+A–)	contemptuous	dogmatic
Bad, impotent, active (E–P–A+)	panicked	immature
Bad, impotent, passive (E–P–A–)	sad	cowardly

[a]Good examples of emotions with the +–+ pattern are unknown, so I give an approximate example.

TABLE 4.6 Design for Study of Amalgamations, with Both Emotions and Traits

Attribute EPA Pattern	Identity EPA Pattern							
	+++	++–	+–+	+––	–++	–+–	––+	–––
+++	Emotion	Trait	Emotion	Trait	Emotion	Trait	Emotion	Trait
++–	Trait	Emotion	Trait	Emotion	Trait	Emotion	Trait	Emotion
+–+	Emotion	Trait	Emotion	Trait	Emotion	Trait	Emotion	Trait
+––	Trait	Emotion	Trait	Emotion	Trait	Emotion	Trait	Emotion
–++	Emotion	Trait	Emotion	Trait	Emotion	Trait	Emotion	Trait
–+–	Trait	Emotion	Trait	Emotion	Trait	Emotion	Trait	Emotion
––+	Emotion	Trait	Emotion	Trait	Emotion	Trait	Emotion	Trait
–––	Trait	Emotion	Trait	Emotion	Trait	Emotion	Trait	Emotion

whether emotion and trait amalgamation processes are equivalent and to study each separately if they are not, although with the low certainty associated with regressions over 32 cases for each type of modifier.

The set of attribute–identity combinations defined by crossing all EPA patterns for a given kind of attribute with all EPA patterns of identities, or the mixed combinations defined in Tables 4.5 and 4.6, have excellent characteristics as bases for regression analyses. In either case, first-order predictors and second-order interactions are all orthogonal, providing optimal efficiency in regressions. (In practice, predictors will have some small correlations because real-life attributes and identities do not implement the designs perfectly.)

Table 4.7 first lists the 32 emotion–identity patterns indicated in Table 4.6, then the 32 patterns identified by the word *Trait*. The symbol for attributes is *S* rather than *A* to avoid confusion with the symbol for Activity. An emotion–identity and a trait–identity example are given for each pattern.

A 64-item study could be assembled from the 32 emotion–identity elements in the top half of Table 4.7, plus the 32 trait–identity elements in the bottom

TABLE 4.7 Attribute (S) with Identity (I) Patterns Defined by Table 4.5, with Examples

Pattern	Emotion Example	Trait Example
	From Emotion Cells in Table 4.6	
S+++, I+++	an elated athlete	a competitive athlete
S+−+, I+++	an emotional teammate	a feminine teammate
S−++, I+++	an enraged athlete	a quarrelsome athlete
S−−+, I+++	an agitated teammate	a conceited teammate
S++−, I++−	a calm grandparent	a sympathetic priest
S+−−, I++−	a serene priest	a cautious grandparent
S−+−, I++−	a displeased grandparent	a vengeful priest
S−−−, I++−	a dejected priest	a lazy grandparent
S+++, I+−+	a gleeful child	a competitive child
S+−+, I+−+	an emotional teenager	an idealistic teenager
S−++, I+−+	an enraged child	a quarrelsome child
S−−+, I+−+	a panicked teenager	a conceited teenager
S++−, I+−−	a calm retiree	a sympathetic old-timer
S+−−, I+−−	a nostalgic old-timer	a cautious retiree
S−+−, I+−−	a displeased retiree	a vengeful old-timer
S−−−, I+−−	a sad old-timer	a lazy retiree
S+++, I−++	an elated bully	a competitive bully
S+−+, I−++	an emotional outlaw	an idealistic outlaw
S−++, I−++	an enraged bully	a ruthless bully
S−−+, I−++	an agitated outlaw	a conceited outlaw
S++−, I−+−	a calm executioner	a sympathetic executioner
S+−−, I−+−	a serene murderess	a cautious murderess
S−+−, I−+−	a contemptuous executioner	a vengeful executioner
S−−−, I−+−	a sad murderess	a cowardly murderess
S+++, I−−+	an elated brat	a competitive gossip
S+−+, I−−+	a sentimental gossip	a feminine brat
S−++, I−−+	an enraged brat	a quarrelsome gossip
S−−+, I−−+	an agitated gossip	a conceited brat
S++−, I−−−	a calm do-nothing	a modest do-nothing
S+−−, I−−−	a nostalgic victim	an obedient victim
S−+−, I−−−	a contemptuous do-nothing	a vengeful do-nothing
S−−−, I−−−	a dejected victim	a cowardly victim
	From Trait Cells in Table 4.6	
S++−, I+++	a calm teammate	a competitive teammate
S+−−, I+++	a nostalgic athlete	a modest athlete
S−+−, I+++	a displeased teammate	a vengeful teammate
S−−−, I+++	a dejected athlete	a lazy athlete
S+++, I++−	an elated grandparent	a courageous grandparent
S+−+, I++−	a sentimental priest	an idealistic priest
S−++, I++−	an enraged grandparent	a quarrelsome grandparent
S−−+, I++−	an agitated priest	a conceited priest
S++−, I+−+	a calm teenagers	a modest teenager

TABLE 4.7 (*Continued*)

Pattern	Emotion Example	Trait Example
S+−−, I+−+	a serene child	an obedient child
S−+−, I+−+	a contemptuous teenager	a vengeful teenager
S−−−, I+−+	a sad child	a cowardly child
S+++, I+−−	a gleeful old-timer	a courageous old-timer
S+−+, I+−−	a sentimental retiree	an idealistic retiree
S−++, I+−−	an irate old-timer	a quarrelsome old-timer
S−−+, I+−−	a panicked retiree	a conceited retiree
S++−, I−++	a calm outlaw	a sympathetic outlaw
S+−−, I−++	a serene bully	a cautious bully
S−+−, I−++	a displeased outlaw	a vengeful outlaw
S−−−, I−++	a sad bully	a cowardly bully
S+++, I−+−	an elated executioner	a courageous executioner
S+−+, I−+−	an emotional murderess	a feminine murderess
S−++, I−+−	an irate executioner	a ruthless executioner
S−−+, I−+−	an agitated murderess	an conceited murderess
S++−, I−−+	a calm gossip	a sympathetic gossip
S+−−, I−−+	a serene brat	an obedient brat
S−+−, I−−+	a displeased murderess	a vengeful gossip
S−−−, I−−+	a dejected brat	a cowardly brat
S+++, I−−−	an elated victim	a courageous victim
S+−+, I−−−	an emotional do-nothing	an idealistic do-nothing
S−++, I−−−	an irate victim	a quarrelsome victim
S−−+, I−−−	an agitated do-nothing	a conceited do-nothing

half. A 128-item study could be assembled from the 64 emotion–identity elements in the column of Table 4.7 giving emotion–identity examples (top and bottom halves of the table combined), plus the 64 trait–identity elements in the column giving trait–identity examples.

The 45 identities, emotions, and traits used to construct example combinations are given in Table 4.8. Some illustrative frames for presenting in-context and out-of-context attribute stimuli are displayed in Table 4.9.

TABLE 4.8 Words Used to Construct Examples

EPA Pattern	Identities	Emotions	Traits
+++	athlete, teammate	elated, gleeful	competitive, courageous
++−	grandparent, priest	calm	modest, sympathetic
+−+	child, teenager	emotional, sentimental	feminine, idealistic
+−−	old-timer, retiree	nostalgic, serene	cautious, obedient
−++	bully, outlaw	enraged, irate	ruthless, quarrelsome
−+−	executioner, murderess	displeased, contemptuous	vengeful
−−+	brat, gossip	agitated, panicked	conceited
−−−	do-nothing, victim	dejected, sad	cowardly, lazy

TABLE 4.9 Frames for Presenting Stimuli When Studying Impressions from Events

An elated athlete
is

A competitive athlete
is

Feeling elated
is

Being competitive
is

5 Errors in Surveys

A major intersection between cultural surveys and traditional survey research arises out of concern with errors arising in people's responses to queries. In this chapter we consider the kinds of errors that occur in surveys, how each kind of error manifests itself in assessments of cultural sentiments, and the extent to which contemporary sentiment measurement procedures provide safeguards with regard to that type of error. Then I develop a measurement model specifically adapted to the problem of assessing cultural sentiments. The model extends essential ideas of the culture-as-consensus paradigm (Romney 1999; Romney, Batchelder, and Weller, 1987; Romney, Weller, and Batchelder 1986).

A number of recent books focus on errors in social surveys (e.g., Alwin 2007; Biemer et al. 2004; Tourangeau, Rips, and Rasinski 2000; Weisberg 2005). Alwin (2007) and Weisberg (2005) distinguish several basic kinds of errors. *Coverage errors* occur when a sampling frame misses parts of a target population of potential respondents. *Sampling errors* arise when the sample of respondents obtained misrepresents the target population. *Nonresponse errors* occur when respondents do not answer a question adequately, because they skip the item or they choose a don't-know option. *Measurement errors* relate to various ways that an answer gets distorted, either by the respondent or by an interviewer. Weisberg also allows for a fifth type, *postsurvey errors*, arising from glitches in the processing and analysis of survey data. This classification of survey errors provides a useful framework for considering the kinds of problems that arise in measuring cultural sentiments.

A measuring instrument used in contemporary surveys of cultural sentiments is described in detail in Chapter 2. It is worth reviewing its features briefly, since the measurement characteristics of this instrument are considered throughout the rest of this chapter. A Java program manages a computer-assisted, self-administered interview. The program begins the rating session by presenting an interactive tutorial on how to record feelings about things on bipolar graphic rating scales assessing three dimensions of affective meaning. Then the program initiates ratings, randomly selecting a stimulus from the set of stimuli to be rated. Sets of adjectives defining one of the affective dimensions are selected randomly, and the left–right orientation of adjectives on the scale is randomized. The respondent rates the stimulus on the scale by dragging a pointer along the scale to a desired position, and then clicking a Save button. The program stores the graphic position of the pointer as a numeric measurement, and presents the next scale, going on to the next stimulus after all three ratings have been made for the current stimulus. The respondent can bypass unfamiliar stimuli by clicking a Skip button, and the respondent can modify previous ratings by clicking a Change button. After a few stimuli have been rated, a line appears at the bottom of the screen, saying "Left to do:" followed by the number of stimuli the respondent still has to rate. When all stimuli have been rated, the program sends the data to a central depository via the Internet.

5.1 COVERAGE ERRORS

A survey has coverage errors when its sampled population is different from the target population. For example, a telephone survey that aims to represent the U.S. population has coverage error since it cannot reach people who are without telephones.

Sentiment measurements are obtained to describe a culture rather than a demographic population. Thus, the target population for a sentiment survey is the set of persons who are applying and reproducing the culture in their everyday lives. The population of people reproducing a culture is not coextensive with the population of a political unit. The United States, for instance, consists of multiple subpopulations associated with different cultures. In some cases the patterns of sentiments within these different cultures are as different from one another as the patterns of different nations, as Sewell and Heise (2009) found for inner-city blacks versus whites.

A focus on behavior settings is commonplace in sociological ethnology where researchers routinely frequent such sites as a factory, a club, or a street corner where culture is being enacted (Herman-Kinney and Verschaeve 2003). Studies involving sentiment measurements have also used this approach. Azar and Lerner (1981) obtained ratings from professionals in the State Department, consulting firms, and political science departments to assess the sentiments that sustain the culture of international relations. Most studies by researchers working with affect control theory employed students in universities to assess the middle-class sentiments sustaining social institutions (e.g., Heise 1979; MacKinnon 1994; Smith, Ike, and Yeng 2002; Smith-Lovin and Heise 1988). Black culture has been assessed from ratings by boys in Chicago street-corner gangs (Gordon et al. 1963) and in center-city and south-side high schools in Chicago (Landis et al. 1976; Sewell and Heise 2009).

Coverage errors arise in culture studies because implemented sampling frames include only a small portion of the settings that are relevant in a study. For example, Azar and Lerner dealt with a subset of offices and people reachable through Edward Azar's acquaintance network (Heise and Lerner 2006,

p. 999), thereby overemphasizing American and Middle Eastern behavior settings and missing settings and agents in South America and non-Muslim Asia and Africa. Students in most affect control theory studies were acquired in educational settings, thereby missing settings and actors in the worlds of work, medicine, politics, and so on. Black culture assessed from ratings by boys in Chicago gangs and high schools missed the contributions of adult and female respondents and respondents in other regions of the United States.

Several problems interfere with broad sampling of behavior settings.

1. *Travel costs.* Many cultures are maintained in settings widely dispersed throughout a nation, a continent, or worldwide. A sampling frame that called for researchers visiting a probability sample of these settings would proliferate travel costs impossibly, especially since culture studies typically are low-budget operations. For reasons of economy, culture researchers typically acquire informants or respondents in one or a few settings, located near the researchers' home bases, and thus produce coverage errors.
2. *Access.* Institutional officials protect their settings from outside intrusions. For instance, officials are likely to eject unauthorized persons handing out questionnaires in a church, school, or business firm. Obtaining respondents in a behavior setting requires permission from institutional officials who control access to the setting, and sometimes officials deny access. Consequently, culture researchers tend to select settings controlled by officials with whom the researchers are friendly, yielding another source of coverage error.
3. *Power issues.* Culture researchers typically seek research participation from all persons currently in a behavior setting rather than sampling the setting's denizens in order to obtain informants or respondents. The official who grants permission to collect data in a setting often facilitates this approach by authoritatively mandating, or at least recommending, participation in the study by the setting's denizens. However, officials' deliveries of setting denizens often do not include participation by the officials themselves, and the nature of power precludes officials from mandating participation by their peers and superiors. Truncation of the upper end of the power continuum among informants and respondents is so common in culture studies that it can be considered a standard coverage error. (The brilliantly implemented study by Berk and Rossi 1977 is a notable exception.)

Studies of cultural sentiments with the computerized instrument described previously have an additional coverage problem: Such studies cannot reach potential respondents who lack access to the Internet or who have insufficient computer skills to use the mouse-driven graphic rating scale. For cultures where most people are computer literate and computer equipped, this presents a negligible coverage problem, just as households without telephones ordinarily are considered negligible in regular survey research. However, lack of computer literacy and computers connected to the Internet might be a problem in assessing sentiments in some cultures within the United States, such as black or Hispanic, and in cultures of third-world nations.

5.2 SAMPLING ERRORS

Sampling error does not arise in culture studies that get the same information from all persons in a setting—just the coverage errors related to visiting few of a culture's behavior settings and omitting the most powerful denizens within sites. Sampling errors arise in sentiment studies, however. Sentiment surveys typically focus on several thousand concepts, or stimuli. The task of rating all stimuli is too large for any one respondent, so stimuli for rating typically are divided into sets that can be completed during a rating session of an hour or less. Each set of stimuli is presented to a different random sample of persons at the behavior setting. Thus, the respondents for any one stimuli set are a probability sample. The standard error of the mean provides information regarding the likelihood that the mean observed for a rating of a given concept on a given scale matches the mean that would have been obtained had everyone in the setting completed the same rating.

Rating statistics do not support statistical inferences about ratings in any population other than the people in a given setting. However, variances of ratings can be decomposed analytically to address various questions about the distribution of cultural sentiments within the studied respondents, as is done in this book.

5.3 NONRESPONSE ERRORS

Nonresponse at the unit level refers to failures to obtain any data from those who were supposed to be respondents. In culture studies, this kind of nonresponse arises when individuals in a setting are not solicited for participation or when a person who has been solicited refuses cooperation. Failures to solicit are most likely to arise when the population of people in a setting is defined comprehensively as all those who frequent the setting during some extended period of time. In such a case, a researcher would have to inhabit the setting for that period of time, or visit the setting frequently, in order to contact and solicit participation from everyone in the population. In the case of sentiment studies, the solicitation issue often is moot because authorities who control access to the setting use their own communication channels to mandate or recommend participation in a study for all persons in the setting.

Whether authorities mandate or recommend participation makes a substantial difference in cooperation rates. Participation that is mandated by an authority leads to virtually 100 percent cooperation, although some respondents provide poor data (as discussed later). However, internal review boards concerned with issues of power increasingly are forbidding this form of solicitation. Participation that is recommended by an authority leads to lower rates of cooperation, with response rates at levels that are familiar to survey researchers. In this case, survey researchers' methods of maximizing cooperation (e.g., see Weisberg 2005, pp. 167–190) may be applicable in ethnological work.

Nonresponse at the item level refers to an item that is unanswered by a respondent who is cooperating in general. This kind of nonresponse is common in studies of cultural sentiments, since such studies routinely allow respondents to skip concepts that are unfamiliar to them. In later chapters we offer several analyses of item nonresponse by respondents.

5.4 MEASUREMENT ERRORS

Weisberg (2005, p. 72) proposed that measurement error "occurs to the extent that the respondents are not providing the answers they should, given the researcher's intentions." Alwin (2007, p. 3) viewed measurement error in the classic psychometric tradition as the "error that occurs when the recorded or observed value is different from the true value of the variable"; and the true value can be estimated as the mean of observations, the errors being random deviations from the mean. Heise (2001b, p. 13508) summarized the perspective of some contemporary measurement theories (e.g., conjoint analysis and structural equation modeling): "Theories set expectations about what should be observed, in contrast to what is observed, and deviation from theoretical expectations is interpreted as measurement error." All three perspectives share a notion that survey respondents' answers would be improved if some unwanted perturbations—the measurement errors—could be removed.

Tourangeau, Rips, and Rasinski (2000) proposed that measurement errors arise during each of the major psychological processes involved in answering a survey question: comprehension, retrieval, judgment, and response. Their categories provide a useful framework for considering the kinds of measurement errors that occur in studies of cultural sentiments.

1. *Comprehension*. Respondents sometimes answer questions based on knowledge that they do not have, so the answers they provide are in error. A question may also introduce errors by incorporating ambiguous words that are interpreted differently by different respondents. Additionally, error arises when respondents—insufficiently motivated to comprehend an item adequately—give any answer that suffices for moving on (Krosnick and Alwin 1987).

The instrument for measuring cultural sentiments used in this book allows respondents to skip a stimulus, the intention in offering this option being explicitly to eliminate errors arising from respondents rating unfamiliar concepts. [Some costs to the respondent are added (i.e., clicking an extra button and waiting) to prevent excessive item nonresponse.] Moreover, each stimulus rated in a sentiment study typically is a word or short word phrase, so overly complex stimuli do not cause comprehension problems. Nevertheless, comprehension problems do arise from three sources: homonymy in stimuli, low cultural inculcation among some respondents, and low motivation among some respondents.

Many words have multiple meanings, so without sufficient context the exact stimulus to be rated may be ambiguous. For example, *doctor* has at least two meanings referring to people, and can also be interpreted as a verb with at least three additional meanings. An early sentiment-measurement study (Heise 1965) tried to disambiguate every stimulus word by usage in a short sentence consisting of function words and generalities, such as "He is a doctor." However, such sentences sometimes skew meanings and may fail to disambiguate adequately; for example, "He is a doctor" skews thought toward males while still not distinguishing between healers and Ph.D.s. Therefore, later studies merely identified nouns with indefinite articles ("a doctor"), interpersonal verbs with an infinitive frame ("to doctor someone"), and personal modifiers with a state-of-being frame ("being intelligent"), except for emotions that used a feeling frame ("feeling joyous"). Norms of word association (Kiss et al. 1973) were presumed to select among multiple meanings of a word; for example the medical sense of a doctor is evoked in most people. In this framework, ratings by respondents who select and rate nonnormative senses of a word produce errors in measuring the sentiment associated with the normative sense. Stimuli with diffuse associations produce more errors of this sort.

Respondents may be imperfectly inculcated into the culture being studied because their socialization is incomplete or because they have been poorly socialized, as in the case of offspring from abusive homes (Weller, Romney, and Orr 1987). Such persons have incomplete understandings of the meanings of some concepts, and errors related to their incomprehension get incorporated into their sentiment ratings. This problem compares with the problem of literacy for self-administered paper-and-pencil questionnaires. This source of measurement error is analyzed in detail in Chapter 6.

Thoroughly unmotivated respondents who have not refused participation in a survey rush through the task quickly. In the case of the sentiment measurement instrument used in this book, that means flicking the pointer randomly on every scale, without attending to the stimulus. Respondents with a modicum of motivation may read the stimulus and flick the pointer in the direction representing their feeling while disregarding precise positioning. Such tactics produce measurement errors in assessing sentiments. Analyses in later chapters consider measurement errors associated with low motivation.

Respondents who are motivated initially can develop ennui after several hundred ratings and may then start engaging in the tactics associated with low motivation. The sentiment-measuring instrument incorporates a number of features known to keep respondents attentive (Schonlau, Fricker, and Elliot 2002). Scales for rating different dimensions of affective meaning are presented one at a time on the computer screen, in random order. The orientation of the scales is randomized, so for example the good side of an evaluation scale sometimes appears on the left side of the screen, sometimes on the right. Progress is reported at the bottom of the screen in a tally of stimuli yet to be rated. Beyond these contrivances to maintain attention, the sentiment-measuring instrument disperses errors due to ennui across stimuli by presenting stimuli in a different random order for each respondent.

2. *Retrieval.* A respondent must search memory for a fact or personal experience to answer many survey questions, and failure to recall the desired information accurately produces error. However, measurements of cultural sentiments involve a different kind of task: reporting affective associations for stimuli presented.

Tourangeau, Rips, and Rasinski (2000, Chapter 6) describe two general types of models specifying how attitudes (i.e., associations on the Evaluation component of sentiments) are retrieved for reporting in survey studies. One class of models proposes that respondents create attitudes on demand by melding various beliefs or values that seem related to the stimulus at the moment of rating. The related considerations assemble feelings whose combination constitutes the attitude that the respondent reports. Another class of models proposes that considerations related to a stimulus are absorbed into the attitude over time as various experiences happen, and rating the stimulus involves reporting the well-formed current attitude in memory that is automatically activated by the stimulus.

With regard to sentiment measurements, the first kind of model applies especially to novel concepts where no cultural sentiment is available for retrieval, or to stimuli that combine elementary concepts in a manner that might be unfamiliar to respondents (e.g., a Christian on welfare). In this case, rating variations will be fairly large across respondents and across time, producing high levels of measurement error, according to research reported in Tourangeau, Rips, and Rasinski (2000, Chapter 6). The second kind of model applies to elementary concepts that are familiar to respondents, overlearned to the point of automaticity in response, and thereby producing little in the way of error from the retrieval process. This model is the one typically used in studies of cultural sentiments, although a model of continuing information integration occasionally is applied with novel or changing concepts (Heise 2006).

3. *Judgment.* A respondent must consolidate retrieved information to answer survey questions that ask for facts or for frequencies and likelihoods of certain kinds of experiences. Some aspects of this process are weighing the accuracy and completeness of retrievals, considering extra information gained during retrieval, and combining items of information into a single conclusion. Less meticulousness in performing these tasks results in measurement errors. Similar judgmental processes apply when reporting sentiments about novel or fairly complex stimuli.

However, judgment is not much of a factor in rating cultural sentiments associated with elementary concepts conveyed in a word or idiomatic phrase, assuming that the sentiment in such cases is well formed and activated automatically by the stimulus. Indeed, instructions for the instrument used in this book advise respondents to "Base your rating on your first impression rather than on logical reasoning" to encourage their accessing sentiments that are available under time pressure (Sanbonmatsu and Fazio 1990) rather than constructing novel sentiments. Actually, even if instructed to work slowly and thoughtfully, respondents typically adopt a rapid pace when rating elementary concepts, after rating a few of the stimuli (Miron 1961).

4. *Response.* Having formed a subjective judgment, a respondent must transform it into one of the permitted behavioral responses in the survey. For example, shrugs, nods, and head-shakes cannot be used other than in face-to-face interviews. With a closed-form question, the respondent must decide which of the options provided best reflects the subjective judgment that has been made. Additionally, social constraints may cause editing of responses. That is, a respondent may sense that an answer best reflecting the judgment that has been formed is too uncivil or socially risky, and therefore gives a safer answer.

The computerized rating scales used in contemporary sentiment measurements require responding via a computer mouse. Respondents necessarily are skilled in using the device, since they must reach the questionnaire by navigating the Internet, but mishaps in eye–hand coordination can still produce errors.

Respondents are presumed to be unfamiliar with rating scales, so the sentiment survey starts off with a tutorial about the rating instrument. Respondent participation in the tutorial assures that the respondent understands that any point along the scale can be used as a response, including the middle position that serves as a neutral position, distinct from a don't-know or skip response. The respondent also learns in the tutorial that the ends of the scale represent polar opposites and that more intense feeling is represented by positioning the pointer farther toward an extreme. Experience indicates that for most respondents, the tutorial eliminates confusion on how to use the instrument.

Computer-assisted self-administered interviews, such as the sentiment-measuring instrument used in this book, are known to reduce editing of responses with socially sensitive questions (Tourangeau, Rips, and Rasinski 2000, pp. 293–295). Even so, responses to some stimuli are more conformist using verbally grounded sentiment measures than using projective techniques of sentiment measurement (Raynolds and Raynolds 1989; Raynolds, Sakamoto, and Raynolds 1988). On the other hand, respondents' editing of their responses to conform with norms arguably reduces measurement error in assessing cultural sentiments. For example, a respondent who feels positively about "a cocaine user" might be socially intimidated into giving a negative rating, reflecting the prevailing negative sentiment within the culture at large.

5.5 OTHER ERRORS

Schonlau, Fricker, and Elliot (2002, pp. 76–79) summarize advantages and disadvantages of Internet surveys, such as sentiment surveys discussed in this book. A key disadvantage of surveying sentiments via the Internet has already been mentioned: lack of coverage of persons without computer skills and equipment.

A technological issue in Internet surveys is that errors can be produced by programming code. The sentiment measurement instrument used in this book remains largely stable from one study to the next, so debugging is not required every time. However, each study using the instrument has questions and stimuli lists that must be translated to Java code, and sometimes errors in the translation go undetected, especially when non-Western languages require use of Unicode identifiers. Another technological issue is the prevention of bungled Internet communications that could result in data loss. The communication channels must be pilot-tested carefully before respondents employ the system.

Internet surveys of cultural sentiments offer some advantages over other modes of gathering such data. The surveys are self-administered on computers, so interviewers do not bias a respondent's answers or misrecord the respondent's answers, as sometimes happens in face-to-face or telephone interviewing. Additionally, data-entry errors introduced after the survey while coding answers from a paper questionnaire and keying the codes to electronic form do not occur, because the sentiment-measuring instrument transforms the respondent's behavioral responses directly into numbers in computer records.

5.6 A SURVEY-OF-CULTURES MODEL

Researchers in the culture-as-consensus research program within psychological anthropology mathematically analyzed informants' reports about cultural norms, providing formulas for estimating the cultural competence of individual informants and for defining the correct (normative) answer for each item presented to respondents (Batchelder and Romney 1988; Karabatsos and Batchelder 2003; Romney, Weller, and Batchelder 1986). Later analyses also developed methods for estimating guessing biases and item difficulties.

The core of the culture-as-consensus research program dealt only with categorical variables: true–false questions, multiple-choice questions, and fill-in-the-blank items. Although the general ideas of the program were extended to rank-order variables (Romney, Batchelder, and Weller 1987) and interval-scale data (Weller 1987), these extensions involved relatively little mathematical analysis and scant attention to the response process. Moreover, the interval-scale model that Weller analyzed with Monte Carlo runs related mostly to a view of attitudes as created on demand rather than as directly accessible from memory, the perspective that best fits the measurement of sentiment norms as conceptualized in this book.

One early publication in the culture-as-consensus research program viewed respondents from a psychometric standpoint, noting that "a group member serves as an informant, tapping group knowledge in exactly the same way that a test item taps subject matter on a test" (1984, p. 61). This kind of understanding provides a framework for generalizing the approach to interval-scale data if linked to the classic psychometric framework provided by Lord and Novick (1968). However, no major effort was made in the culture-as-consensus research program to modify the classic psychometric model for surveys of cultural norms, in order to treat individual differences as a kind of error and to deal explicitly with response errors that arise from low levels of cultural inculcation. I do that in the following sections of this chapter.

My mathematical analysis takes off from Lord and Novick's (1968) masterwork on psychological measurement, especially their Chapter 2 formulation of analyses across persons and analyses across tests, and I use several measurement-theory assumptions that they explicated. In support of later chapters, in this chapter I present a model of continuous-variable indicators of cultural norms and develop equations defining means, variances, and covariances of norm measurements. The model explicitly acknowledges that respondents may be less than perfectly inculcated into the culture of interest, and that respondents' interrespondent variation is a kind of error, along with intrarespondent variation. Concern with individuals' levels of cultural inculcation parallels the concern of anthropologists with informants' cultural competence and reliability (see, e.g., D'Andrade 1995, pp. 212–217; Romney, Weller, and Batchelder 1986).

The path diagram in Figure 5.1, an extension of a model developed by Heise and Bohrnstedt (1970), illuminates the idea of averaging individuals' ratings in order to estimate cultural sentiments. The diagram applies to a specific concept and to either the Evaluation, Potency, or Activity dimensions of sentiments. A cultural norm is displayed as the common source for individuals' sentiments, this representation being a "causal approximation" (Heise 1975, p. 31) to represent overarching correlations among the sentiments of all individuals.[1]

The premises of the model are as follows. As respondents assess their sentiment about a concept and rate the concept on a scale, they:

1. Adopt the normative sentiment about the concept in their culture
2. Adjust the normative sentiment with their unique feelings about the concept based on personal experiences, as represented by the U variables in Figure 5.1

[1]Lines between the U variables in Figure 5.1 indicate correlations in respondents' sentiments beyond those associated with the cultural norm. I include such correlations in the model for two reasons. First, they facilitate the substantive inferences following equation (5.10); second, such correlations within a subset of respondents represent a shared subculture, discussed in Chapter 7.

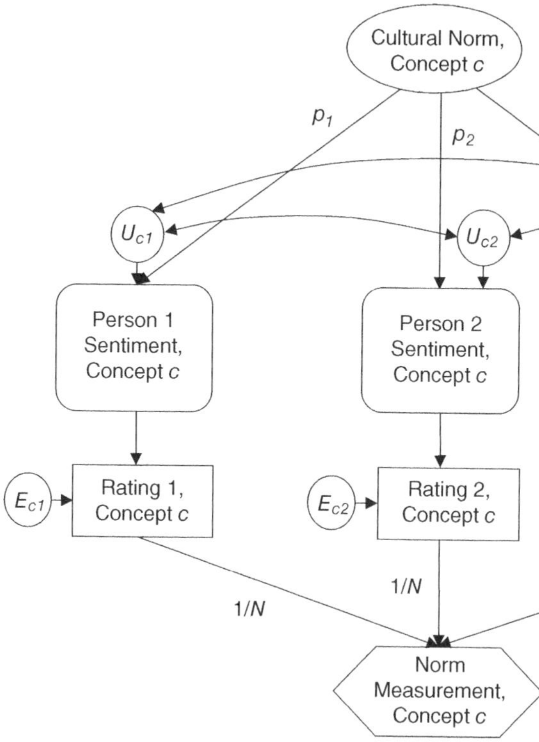

FIGURE 5.1 Causal model of norm influences on respondents, and respondent influences on norm measurement.

3. Sustain some degree of error in transforming their feelings into a rating on a scale, as represented by the *E* variables in Figure 5.1

In this model, a person's sentiment about a concept is a summation of the cultural norm and the person's idiosyncratic feelings about the concept. Such a conception corresponds to the sentiment-formation process described in dynamic terms by Friedkin (1998, p. 24): "At each point in the social influence process, (a) a network 'norm' is formed for each actor that is a weighted average of the influential opinions for the actor and (b) a modified opinion is formed for each actor that is a weighted average of the network 'norm' and the initial opinion of the actor." Holland's (1987, p. 247) interactionist model specifies conditions under which the model would hold: face-to-face interaction among people in stable, nonsuperficial relationships, using group knowledge to which all are equally entitled. My own interpretation goes beyond social networks in viewing the model's preconditions in terms of culture, thereby allowing for inculcation of sentiment norms via language learning and ecological shaping of beliefs and feelings, as well as through direct social influence in groups.

An assumption in derivations below is that some people are uncertain about the cultural norms applying to core concepts, and they convert their uncertainty into tentative ratings pulled toward the zero point of the scale, rating the concepts less extremely as their uncertainty is greater. Although phrased conversely, this is similar to the proposition that more extreme sentiments about a concept emerge from greater involvement with the concept (Liu and Latané 1998; Thomsen, Borgida, and Lavine 1995). For example, an American may think that most Americans greatly value educating someone, but she is not sure because she has paid little attention to the issue, so she tones down her own rating of educating someone to slightly good. The formulation emerges from a study of rating variations by Thomas and Heise (1995), who factor-analyzed ratings of 108 concepts by 326 respondents and found that extremity of responses was the major distinction between respondents with high and low scores on the first principal component, which was "defined mainly by contrasts between the neutral ratings of people in [one group] as opposed to the nonneutral ratings of other respondents, and the items that best define the component are simply those where ratings of other respondents are most extreme" (Thomas and Heise 1995, p. 433). Intensive interviews of a few respondents suggested that those with low extremity of responses were less aligned with cultural norms than those with high extremity of responses; they expressed alienation, confusion about expectations, and they rated nearly every concept as near neutral, reflecting highly attenuated sentiments (Thomas and Heise 1995, p. 434).

Similarly, Rossi and Berk (1985, p. 343) reported that findings from multiple studies of normative consensus "suggest that disputes over norms do not often take the form of opposing normative systems. Normative disagreements instead appear to be more frequently centered around degrees of adherence to an agreed-upon normative system."

Accordingly, I assume that people adopt the norms of their reference groups into their own dispositions, with attenuation when uncertain about the norm. The uncertainty factor is represented by the inculcation weight, p_i, which moderates the impact of the normative cultural sentiment on individual i's personal sentiment about concept c.

A person's rating of a concept reflects the person's sentiment about the concept, but ratings also vary as a function of measurement errors arising from various sources, such as clumsiness in using a computer mouse to position the scale pointer, fuzziness in translating subjective feelings into scale positions, or temporary deflections of a respondent's feelings from the person's stable sentiment. The composite effect from all such factors creates a time-specific deviation of the rating from the person's stable sentiment, and this deviation is interpreted as measurement error.

The average of respondents' ratings, located at the bottom of Figure 5.1, estimates the culturally normative sentiment about a particular concept (represented at the top of Figure 5.1) because the cultural norm affects every respondent who is inculcated into the culture, and therefore each respondent's rating reflects the norm. However, covariation in unique feelings (shown in Figure 5.1 by the curved lines between *U*s) represents shared impacts of unspecified factors, which can bias the norm measurement. Moreover, a person's idiosyncratic feelings and measurement errors add nonnormative constituents to individuals' ratings, and these get carried forward into the average. Thus, the average rating of a concept is an impure measurement of the normative cultural sentiment for that concept.

5.6.1 Definitions and Assumptions

I use a number of terms throughout this chapter and elsewhere in the book, and terse definitions of these are provided below. The following list also includes assumptions that are used in the derivations of this chapter to simplify algebraic expressions.

- *Core concepts.* Core concepts are those understood by every competent, well-inculcated member of the culture. Subscripts representing different concepts, c, go from 1 to M, where M represents the total number of core cultural concepts.
- *Sentiment norm.* A sentiment norm is an affective association to a given concept, shared by members of a culture. The value of the norm for a given concept, W_c, is subscripted to indicate the relevant concept.
- *Individual respondent.* A respondent is a participant in a culture survey. The respondent's sentiment, S_{ci}, is subscripted to indicate which concept and which person is being considered. Subscripts identifying individuals, i, go from 1 to N, the total number of people under consideration.
- *Uniformity of inculcation.* Weller (1987, p. 178) observed that "the sheer magnitude of the 'pool' of cultural knowledge prohibits individual mastery and makes variation in knowledge among individuals inevitable." However, studies of cultural sentiments ordinarily focus on core concepts that are understood by every competent, well-inculcated member of the culture. Levels of inculcation are assumed to be the same across such core concepts, even though people vary in their overall levels of inculcation of all concepts.
- *Inculcation.* An inculcation weight, p_i, indicates how cultural norms affect each person's sentiments. Since the weight is the same for all core concepts, subscripting identifies just the person involved, not the concept under consideration. An inculcation weight of zero represents a complete lack of inculcation into a given culture, meaning that the person's sentiments are experience-based or derived from a dissimilar culture. (For the weight to represent inculcation in this way, the neutral or indifference point in the scale metric must correspond to zero. Such a metric can always be achieved through appropriate coding of rating-scale positions, and such a metric is used routinely in sentiment measurements with

bipolar rating scales such as that shown in Figure 5.1.) A value between zero and one means that the person's sentiments reflect cultural norms but with some degree of attenuation. Conceivably a person could have an inculcation weight above 1.0, meaning that the person's sentiments exaggerate cultural norms. Also, a person conceivably could have a negative inculcation weight, meaning that the person expresses sentiments that systematically oppose cultural norms.

- *Idiosyncrasy.* A person's unique feeling about a concept based on personal experience, U, is measured as a deviation from the culturally inculcated sentiment. Subscripting indicates the relevant concept and person. I assume that any across-individuals bias in idiosyncratic sentiments about a concept gets absorbed into the cultural norm. Therefore, idiosyncratic sentiments about a concept sum to zero across individuals, and the culturally normative sentiment can be viewed as the average of individuals' sentiments. On the other hand, idiosyncratic sentiments may correlate across concepts, as when a subculture provides a unique sentiment for its members.
- *Rating.* A person's rating of a concept, R_{ci}, is subscripted to indicate the relevant concept and person.
- *Measurement error.* Measurement errors are variations in ratings that are not attributable to norms or to individuals' stable deviations from norms. The error, E_{ci}, involved in a person's sentiment rating is subscripted to indicate the relevant concept and person. Time of rating also will be subscripted where that is a factor under consideration. In line with the classic theory of errors (Lord and Novick 1968), measurement errors by individual i sum to zero across different times of measurement and across concepts. Additionally, measurement errors for concept c sum to zero across individuals. Measurement errors are uncorrelated across individuals, times, and concepts; and measurement errors are uncorrelated with either the idiosyncratic or cultural components of sentiments being measured.

5.6.2 Basic Structural Equations

Reading from Figure 5.1, the basic equation defining the composition of a person's sentiment regarding a concept is

$$S_{ci} = p_i W_c + U_{ci} \tag{5.1}$$

where S_{ci} is the sentiment, c is an identification number for the concept being rated, i is an identification number for the person, W_c is a constant representing the normative sentiment about concept c in the given culture, and U_{ci} is the unique component of individual $_i$'s sentiment. The p_i coefficient represents respondent i's level of inculcation into the general culture, which is the same for all concepts.

The basic equation defining the production of a rating is

$$R_{ci} = S_{ci} + E_{ci} = p_i W_c + U_{ci} + E_{ci} \tag{5.2}$$

where R_{ci} is individual i's rating of concept c, and E_{ci} is individual i's error in rating that concept. The final expression in equation (5.2) substitutes equation (5.1) for S_{ci} in order to define a rating in terms of a normative sentiment, an idiosyncratic sentiment deviating from the norm, and measurement error.

In a project designed to explain correlations among different semantic differential scales, Kahneman (1963) presented a similar model for semantic differential ratings, initially without the inculcation coefficient, and later, adding a coefficient similar to p_i (in his equation 3) to account for empirical findings regarding concept-scale interaction. Kahneman called his multiplier coefficient a "constant of exaggeration," and he viewed the constant as a kind of psychological trait varying from one respondent to another.

The equation defining a norm measurement based on the average of respondent ratings is

$$\mu_{R_c} = \frac{1}{N}\sum_{i=1}^{N} R_{ci} = \frac{1}{N}\sum_{i=1}^{N}(p_i W_c + U_{ci} + E_{ci}) \tag{5.3}$$

where μ_{R_c} is the statistical estimate of the cultural norm regarding concept c based on individuals' ratings, R.

5.7 STATISTICS

I now develop statistical implications of the measurement model. For this task I convert from summations to the algebra of expectations, as is conventional in statistical analyses. I employ the conventions that Lord and Novick (1968) used for the algebra of expectations (see also Winkler and Hays 1970, pp. 144–147).

Converting to the algebra of expectations, equation (5.3) becomes

$$\mu_{R_c} = \mathrm{E}(R_{ci}) = \mathrm{E}(p_i W_c + U_{ci} + E_{ci}) \tag{5.4}$$

Next I consider the means, variances, and covariances of ratings where some individuals are incognizant of cultural sentiments to a degree.

5.7.1 Analysis Across Respondents

With some respondents incorporating culture less than perfectly, the mean of ratings across respondents does not estimate the cultural norm exactly, as demonstrated in the following continuation of equation (5.4). The result is simplified using the assumptions that idiosyncrasies and errors have means of zero across individuals, as stated in Section 5.6.1:

$$\mu_{R_c} = \mathrm{E}(p_* W_c + U_{c*} + E_{c*}) = \mu_p W_c \tag{5.5}$$

Equation (5.5) shows that the mean of ratings may be an attenuated estimator of the cultural norm, in that it incorporates the mean of the weights for all respondents, and that mean could be a number less than 1 when some respondents are uncertain about the norm. (Norm exaggerators could balance attenuators, but inadequately socialized persons probably are more common than perfectly socialized exaggerators.)

The variance of individuals' ratings of a concept is the expected value of the ratings' squared deviations from the mean rating. In the following derivation it is assumed that respondents' amounts of culture inculcation are uncorrelated with their feelings about a concept based on personal experience, and with measurement errors. Also, idiosyncratic sentiments and errors are uncorrelated, and both have a mean of zero across individuals.

$$\begin{aligned}
\sigma_{R_c}^2 &= \mathrm{E}\left((R_{c*} - \mu_{R_c})^2\right) \\
&= \mathrm{E}\left((p_* W_c + U_{c*} + E_{c*} - \mu_p W_c)^2\right) \\
&= \mathrm{E}\left((W_c (p_* - \mu_p) + U_{c*} + E_{c*})^2\right) \\
&= \mathrm{E}\left(W_c^2 (p_* - \mu_p)^2 + 2W_c (p_* - \mu_p)(U_{c*} + E_{c*}) + U_{c*}^2 + 2U_{c*}E_{c*} + E_{c*}^2\right) \\
&= W_c^2 \sigma_p^2 + \sigma_{U_c}^2 + \sigma_E^2
\end{aligned} \tag{5.6}$$

Thus, aside from the contributions of idiosyncratic variance and error variance, the rating variance for a concept gets bigger with more variance in inculcation weights, such as when half the respondents are excellent representatives of a culture and the rest rate the concept as neutral because they know nothing of the culture. The variable-inculcation effect is especially large when an extreme cultural norm magnifies the difference between culturally accurate ratings and neutral ratings.

A similar factor affects covariances of ratings for different concepts, *c* and *d*. The following derivation makes use of the assumptions that a person's inculcation weight is the same for all concepts, and idiosyncratic variance and error variance are uncorrelated among themselves or with inculcation of culture.

$$\begin{aligned}
\sigma_{R_c R_d} &= \mathrm{E}((R_{c*} - \mu_{R_c})(R_{d*} - \mu_{R_d})) \\
&= \mathrm{E}((W_c(p_* - \mu_p) + U_{c*} + E_{c*})(W_d(p_* - \mu_p) + U_{d*} + E_{d*})) \\
&= W_c W_d \sigma_p^2 + \sigma_{U_c U_d}
\end{aligned} \tag{5.7}$$

Equation (5.7) shows that when some respondents have inculcated cultural norms imperfectly, the covariance of ratings for two concepts is an impure measure of similarity in idiosyncratic sentiments about the concepts because the covariance also reflects levels of uncertainty about norms and the extremity of the norms.[2] A covariance may become large just because the culture inculcation of respondents is highly variable, especially when the normative sentiments for the concepts are extreme.

5.7.2 Analysis Across Concepts

Lord and Novick (1968, Figure 2.5.2) note that analyses of different measurements can be conducted with individual respondents as the unit of analyses, or that analyses of respondents can be conducted with different measurements as the unit of analyses (the latter is called the *Q-technique* in the factor analytic literature; see Rummel 1970, pp. 192–202). This section deals with analyses of respondents based on computations across the respondents' ratings of multiple concepts.

As shown in equation (5.8), the expected value of a respondent's ratings across all concepts cannot be assumed to be zero, and cultural inculcation affects analyses of individuals based on their ratings of multiple concepts. Assuming that measurement errors have a mean of zero,

$$\begin{aligned}\mu_{R_i} &= \mathrm{E}(p_i W_* + U_{*i} + E_{*i}) \\ &= p_i \mu_W + \mu_{U_i}\end{aligned} \tag{5.8}$$

Equation (5.8) shows that the mean of a person's ratings of a set of concepts is centered at the mean of cultural sentiments for those concepts, μ_W, scaled by the person's inculcation parameter. Additionally, the person's mean rating reflects the person's bias with regard to the concepts: the average of the person's deviations from norms, which might be positive (e.g., an evaluative Pollyanna bias) or negative (e.g., a cynicism bias). Multiple individuals having parallel biases creates covariance between idiosyncratic sentiments for pairs

[2]The presence of respondents with imperfect inculcation of sentiment norms also may bias the covariance of a sentiment measure with another variable, Y.

$$\begin{aligned}\sigma_{R_cY} &= \mathrm{E}((R_{c*} - \mu_{R_c})(Y_* - \mu_{Y_*})) \\ &= \mathrm{E}((W_{0c}(f_{0c*} - \mu_{f_{0c}}) + \cdots + W_{kc}(f_{kc*} - \mu_{f_{kc}}) + W_{Lc}(f_{Lc*} - \mu_{f_{Lc}}) + U_{c*} + E_{c*})(Y_* - \mu_{Y*})) \\ &= W_{0c}\sigma_{f_0Y} + \cdots + W_{kc}\sigma_{f_{kc}Y} + W_{Lc}\sigma_{f_{Lc}Y} + \sigma_{U_cY}\end{aligned}$$

The last expression shows that the covariance of Y with sentiment ratings reflects the relation between Y and individual variations in sentiment, as a researcher might hope, if Y is unrelated to levels of inculcation or if the sentiment norm is zero. But if Y does predict levels of cultural inculcation to some degree, the covariance between Y and sentiment ratings includes a component based on the sentiment norm. Consider, for example, a variable such as age, which relates to inculcation; age may correlate with attitudes about evaluatively extreme concepts just because younger persons have not yet inculcated the norms, even if age is nonpredictive of experience-based variations in attitudes.

of concepts. Since idiosyncratic sentiments sum to zero for each concept, idiosyncratic sentiments of other people compensate for a person's bias.

The variance of a respondent's ratings over all concepts is also affected by the respondent's level of cultural inculcation as shown in equation (5.9). Terms are dropped in the following derivation on the assumptions that individual variations in sentiments are uncorrelated with cultural norms; measurement errors are uncorrelated with cultural norms, individual variations are uncorrelated with measurement errors, and the expected value of a person's measurement errors is zero over concepts:

$$\begin{aligned}
\sigma_{R_i}^2 &= \mathrm{E}\left((R_{*i} - \mu_{R_i})^2\right) \\
&= \mathrm{E}\left((p_i W_* + U_{*i} + E_{*i} - p_i \mu_W - \mu_{U_i})^2\right) \\
&= \mathrm{E}\left(((p_i (W_* - \mu_W) + (U_{*i} - \mu_{U_i}) + E_{*i})^2\right) \\
&= p_i^2 \sigma_W^2 + \sigma_{U_i}^2 + \sigma_{E_i}^2
\end{aligned} \tag{5.9}$$

Equation (5.9) shows that low levels of cultural inculcation mute the contribution of culture to the variability of a respondent's ratings. At the extreme of zero cultural inculcation, the variability of ratings depends only on individual variance and measurement-error variance. From another perspective, the implication of equation (5.9) is that inculcation of a culture increases the variability of a person's sentiments and can produce a variability in sentiments that might otherwise be achieved only through diverse personal experiences.

Equation (5.10) shows how the covariance of two respondents' ratings is influenced by cultural inculcation. Measurement errors drop out as a consideration since errors by one respondent are uncorrelated with the errors of another.

$$\begin{aligned}
\sigma_{R_i R_j} &= \mathrm{E}\left((R_{*i} - \mu_{R_i})(R_{*j} - \mu_{R_j})\right) \\
&= \mathrm{E}\left((p_i W_* + U_{*i} + E_{*i} - p_i \mu_W - \mu_{U_i})(p_j W_* + U_{*j} + E_{*j} - p_j \mu_W - \mu_{U_j})\right) \\
&= \mathrm{E}\left((p_i (W_* - \mu_W) + (U_{*i} - \mu_{U_i}) + E_{*i})(p_j (W_* - \mu_W) + (U_{*j} - \mu_{U_j}) + E_{*j})\right) \\
&= p_i p_j \sigma_W^2 + \sigma_{U_i U_j}
\end{aligned} \tag{5.10}$$

Equation (5.10) reveals that the covariance of ratings by two respondents across concepts consists of the variance of the cultural norms for those concepts—which is constant for all respondent pairs—scaled by the product of each respondent's inculcation parameter, plus the covariance of the respondents' idiosyncratic deviations from the norms.

One substantive implication of equation (5.10) becomes evident by assuming that one (or both) of two individuals has essentially no cultural inculcation. Then sentiments of the two people will co-vary only to the extent that their sentiment-producing personal experiences are similar (e.g., through direct association with one another or through experience in similar environments).

A second implication is the complement of this. The sentiments of two people may co-vary because of their high levels of cultural inculcation, even if they have no sentiment-producing personal experiences in common.

5.7.3 Overall Mean

An expression for the overall mean of ratings across both individuals and concepts will be needed later. Taking the expectation, across concepts, of means for concepts, computed across individuals, as defined in equation (5.5), we get the following.

$$\begin{aligned} E_c(\mu_{R_c}) &= E_c(\mu_p W_c) \\ &= \mu_p \mu_W \end{aligned} \tag{5.11}$$

To check, take the expectation across individuals of means for individuals computed across concepts as defined in equation (5.8), and recall that idiosyncratic sentiments about a concept are assumed to sum to zero across individuals.

$$\begin{aligned} E_i(\mu_{R_i}) &= E_i(p_i \mu_W + \mu_{U_i}) \\ &= \mu_p \mu_W + E_i(E_c(U_{ci})) \\ &= \mu_p \mu_W + E_c(E_i(U_{ci})) \\ &= \mu_p \mu_W + E_c(\mu_{U_c}) \\ &= \mu_p \mu_W \end{aligned} \tag{5.12}$$

5.8 INCULCATION INDEX

Suppose that individual *i*'s sentiments about core concepts could be plotted against cultural norms. Then the regression line through the distribution of points would have a slope of p_i. Of course, this method of estimating p_i is impossible since the values of cultural sentiments are unknown. However, a ranking of respondents corresponding to their inculcation weights, p_i, can be obtained by regressing respondent ratings on the best estimates of cultural norms—the mean ratings by all respondents.

The regression coefficient for predicting one variable from another is the ratio of the variables' covariance divided by the variance of the predictor (Van de Geer 1971, p. 95). Thus, the regression coefficient for predicting a respondent's ratings from the means of all respondent ratings can be decomposed as follows:

$$b_{R_i \mu_{RC}} = \frac{\sigma_{R_i \mu_{RC}}}{\sigma^2_{\mu_{RC}}} = \frac{E_c\left((R_{*i} - \mu_{R_i})(\mu_{R_C} - \mu_{\mu_{RC}})\right)}{E_c\left((\mu_{R_C} - \mu_{\mu_{RC}})^2\right)} \tag{5.13}$$

Equation (5.13) refers to means of ratings computed across individuals and across concepts. The two are distinguished by subscripting a mean across individuals with c to show that it is for a particular concept, and subscripting a mean across concepts with i to show it is for a particular person. Since the subscripts have to be retained for clarity, I show explicitly which index is applicable in expectations rather than conveying this information implicitly through the use of a star subscript.

Equation (5.13) shows that this regression coefficient is built from four elements. *The ratings of concepts by respondent I*; the composition of this element is given in equation (2). *The mean of respondent i's ratings*; the composition of this mean is given in equation (5.8). *The mean of all respondents' ratings of a given concept*; the composition of this mean is given in equation (5.5). *The mean of the mean ratings of concepts*; this is given in equation (5.11).

Substituting the decompositions of the elements into equation (5.13) gives

$$\begin{aligned} b_{R_i \mu_{RC}} &= \frac{\mathrm{E}((p_i W_* + U_{*i} + E_{*i} - p_i \mu_W - \mu_{U_i})(\mu_p W_* - \mu_p \mu_W))}{\mathrm{E}\left((\mu_p W_* - \mu_p \mu_W)^2\right)} \\ &= \frac{\mathrm{E}((p_i (W_* - \mu_W) + (U_{*i} - \mu_{U_i}) + E_{*i})(\mu_p (W_* - \mu_W)))}{\mathrm{E}\left((\mu_p^2 (W_* - \mu_W))^2\right)} \\ &= \frac{\mathrm{E}\left(p_i \mu_p (W_* - \mu_W)^2 + \mu_p (W_* - \mu_W)(U_{*i} - \mu_{U_i}) + \mu_p (W_* - \mu_W) E_{*i}\right)}{\mathrm{E}\left((\mu_p^2 (W_* - \mu_W))^2\right)} \end{aligned} \tag{5.14}$$

As before, it is assumed that normative sentiments do not co-vary with a person's unique variations or with a person's rating errors. Thus, parts of the numerator in equation (5.14) vanish because they involve such covariances, and the expression reduces as follows:

$$\begin{aligned} b_{R_i \mu_{RC}} &= \frac{p_i \mu_p \sigma_W^2}{\mu_p^2 \sigma_W^2} \\ &= \frac{p_i}{\mu_p} \end{aligned} \tag{5.15}$$

Equation (5.15) provides a formula for estimating the relative levels of respondents' inculcation weights. The expected value of the weights within a study is 1.0, as we find empirically in Chapter 6. Estimates are not comparable across studies because they depend on the average quality of respondents in each study, which cannot be determined empirically. On the other hand, if respondents in different studies are drawn randomly from a single pool of raters, one might assume that the average quality of raters is about the same, and inculcation weights computed with equation (5.15) are comparable across studies. In Chapter 6 we assume this kind of comparability across studies to examine correlates of inculcation.

5.9 COMMONALITY INDEX

By a standard statistical formula, the correlation coefficient relating respondent *i*'s ratings of concepts to mean ratings of the concepts by all respondents is as follows[3]:

$$r_{R_i\mu_{RC}} = \frac{\sigma_{R_i\mu_{RC}}}{\sigma_{\mu_{RC}}\sigma_{R_i}} \tag{5.16}$$

The numerator of equation (5.15) shows the model-generated covariance, and the denominator of equation (5.15) shows the model-generated variance of mean ratings. Substituting these terms into equation (5.16) gives

$$\begin{aligned} r_{R_i\mu_{RC}} &= \frac{p_i\mu_p\sigma_W^2}{(\mu_p\sigma_W)\sigma_{R_i}} \\ &= \frac{p_i\sigma_W}{\sigma_{R_i}} \end{aligned} \tag{5.17}$$

The standard deviation of respondent *i*'s ratings in equation (5.17) can be expanded according to equation (5.9), yielding

$$r_{R_i\mu_{RC}} = \frac{p_i\sigma_W}{\sqrt{p_i^2\sigma_W^2 + \sigma_{U_i}^2 + \sigma_{E_i}^2}} \tag{5.18}$$

Equation (5.18) demonstrates that if respondent *i*'s ratings had no unique variance and no error variance, the respondent's ratings would be generated solely from the cultural norms and would correlate 1.0 with the cultural norms (if the inculcation index is above zero). On the other hand, if the respondent's inculcation weight is zero, the cultural norms have no impact on the respondent's ratings, all of the ratings variance would be unique or error based, and the correlation would be zero. A correlation between zero and one reflects some unknown combination of attenuated inculcation, unique sentiments derived from personal experiences, and measurement errors.

The square of the individual-to-total correlation estimates the proportion of the variance in a respondent's ratings that is associated with the general sentiment norms. One minus the square of the individual-to-total correlation provides an estimate of the proportion of the variance in a respondent's

[3]This is comparable to an item-to-total correlation in psychometric theory: that is, the correlation of responses to one item in a scale with summed responses to all items in the scale. Sometimes item-to-total correlations are computed as part correlations, with the variance of the focal item being removed from the total scale variance. I do not complicate derivations here with such a correction because number of respondents in sentiment studies is usually 30 or more (comparable to a scale of 30 or more items), so the correction provided by a part correlation is tiny. However, my empirical estimates of commonality in Chapter 7 make the correction.

sentiments that is not associated with the general sentiment norms. The non-normative component includes both unique variations in sentiments and variations associated with measurement errors:

$$
\begin{aligned}
1-r^2_{R_i\mu_{RC}} &= 1-\frac{p_i^{\,2}\sigma^2_W}{p_i^{\,2}\sigma^2_W+\sigma^2_{U_i}+\sigma^2_{E_i}} \\
&= \frac{p_i^{\,2}\sigma^2_W+\sigma^2_{U_i}+\sigma^2_{E_i}-p_i^{\,2}\sigma^2_W}{p_i^{\,2}\sigma^2_W+\sigma^2_{U_i}+\sigma^2_{E_i}} \\
&= \frac{\sigma^2_{U_i}+\sigma^2_{E_i}}{p_i^{\,2}\sigma^2_W+\sigma^2_{U_i}+\sigma^2_{E_i}}
\end{aligned} \tag{5.19}
$$

I will call the individual-to-total correlation defined in equation (5.17) a *commonality index*. In a later chapter, I employ the commonality index as a dependent variable in analyses of variance and regression analyses. To normalize the distribution of this correlation coefficient, I use Fisher's r-to-z transformation (Winkler and Hays 1970, p. 653), times 2:

$$
\begin{aligned}
z_{R_i\mu_{RC}} &= 2\left(\frac{1}{2}\ln\frac{1+r_{R_i\mu_{RC}}}{1-r_{R_i\mu_{RC}}}\right) \\
&= \ln\frac{1+r_{R_i\mu_{RC}}}{1-r_{R_i\mu_{RC}}}
\end{aligned} \tag{5.20}
$$

5.10 VARIANCE COMPONENTS

Reinterpreting equation (5.6), the variance of ratings of a given concept at a given time is the sum of individual variance and error variance:

$$
\begin{aligned}
\sigma^2_{R_c} &= \left(W_c^2\sigma^2_p+\sigma^2_{U_c}\right)+\sigma^2_E \\
&= \sigma^2_c+\sigma^2_E
\end{aligned} \tag{5.21}
$$

Individual variance consists of individual variance in inculcation of the concept sentiment, weighted by the square of the cultural sentiment for the concept, plus unique variance arising from individuals' differing experiences with the concept. A reliability coefficient for a measurement estimates the proportion of total variance that is meaningful. Traditionally, the meaningful variance is interpreted as the individual variance, and the total variance is the individual variance plus the error variance. Thereby the reliability of a sentiment measurement of concept c is

$$
\begin{aligned}
r_{cc} &= \frac{\sigma^2_c}{\sigma^2_{R_c}} \\
&= \frac{\sigma^2_c}{\sigma^2_c+\sigma^2_E}
\end{aligned} \tag{5.22}
$$

An ideal group of culture informants all would be well and equally socialized into their culture. Thereby they would have the same sentiments, and their meaningful individual variance, σ_c^2, would be zero. Consequently, the reliability of assessing norm *c* from their ratings is zero, according to equation (5.22)!

When measuring cultural sentiments, the meaningful variance instead could be interpreted as the variance of sentiments across concepts. Equation (5.9) shows the composition of individual *i*'s ratings of a variety of concepts, and this variance can be divided into a cultural component and a culturally irrelevant component, which consists of the variance of individual *i*'s idiosyncratic sentiments about the concepts plus the variance of measurement errors.

$$\begin{aligned}\sigma_{R_i}^2 &= p_i^2\sigma_W^2 + \sigma_{U_i}^2 + \sigma_{E_i}^2 \\ &= p_i^2\sigma_W^2 + \left(\sigma_{U_i}^2 + \sigma_{E_i}^2\right)\end{aligned} \tag{5.23}$$

Now the reliability coefficient for individual *i* is the meaningful variance over the total variance:

$$r_{ii} = \frac{p_i^2\sigma_W^2}{p_i^2\sigma_W^2 + \sigma_{U_i}^2 + \sigma_{E_i}^2} \tag{5.24}$$

This is the square of the individual-to-total correlation, or commonality index, given in equation (5.18). Since the commonality index is a simple transformation of a reliability coefficient, the commonality index can be construed as an index of the reliability of individual *i*'s ratings of the given set of concepts.

5.10.1 Cultural Variance

A standard theorem shows how variance of one variable may be partitioned, conditional on values of another variable (Lord and Novick 1968, Theorem 2.6.2):

$$\sigma_R^2 = \mathrm{E}_c\left(\sigma_{R|c}^2\right) + \sigma_{\mathrm{E}(R|c)}^2 \tag{5.25}$$

As Lord and Novick (1968, p. 35) said: "In effect this formula is a generalization of the usual analysis of variance breakdown of total variance into the sum of (1) average within-class variance and (2) among-class variance." Translating to the symbols I have been using, and substituting quantities according to equations (5.5), (5.6), and (5.11), yields

$$\begin{aligned}\sigma_R^2 &= \mathrm{E}_c\left(\sigma_{R_c}^2\right) + \mathrm{E}_c\left(\left(\mu_{R_c} - \mu_R\right)^2\right) \\ &= \mathrm{E}_c\left(W_c^2\sigma_p^2 + \sigma_{U_c}^2 + \sigma_E^2\right) + \mathrm{E}_c\left(\left(\mu_p W_c - \mu_p \mu_W\right)^2\right) \\ &= \sigma_p^2\mathrm{E}_c\left(W_c^2\right) + \mathrm{E}_c\left(\sigma_{U_c}^2\right) + \sigma_E^2 + \mu_p^2\sigma_W^2 \\ &= \sigma_p^2\mu_{W^2} + \mu_{\sigma_U^2} + \sigma_E^2 + \mu_p^2\sigma_W^2\end{aligned} \tag{5.26}$$

Equation (5.27) gives the same result partitioned with parentheses to show the average within-concept variance (first term) and among-concept variance (second term):

$$\sigma_R^2 = \left(\sigma_p^2 \mu_{W^2} + \mu_{\sigma_U^2} + \sigma_E^2\right) + \mu_p^2 \sigma_W^2 \tag{5.27}$$

The proportion of rating variance associated with among-concept variations then is

$$\frac{\sigma_{\mu_{R_c}}^2}{\sigma_R^2} = \frac{\mu_p^2 \sigma_W^2}{\sigma_p^2 \mu_{W^2} + \mu_{\sigma_U^2} + \sigma_E^2 + \mu_p^2 \sigma_W^2} \tag{5.28}$$

This can be interpreted as the reliability coefficient of the measuring instrument with given concepts and individuals:

$$r_{\{i\ldots,c\ldots\}\{i\ldots,c\ldots\}} = \frac{\mu_p^2 \sigma_W^2}{\sigma_p^2 \mu_{W^2} + \mu_{\sigma_U^2} + \sigma_E^2 + \mu_p^2 \sigma_W^2} \tag{5.29}$$

If all people had perfect inculcation, the mean of the inculcation weights would be 1.0, the variance of inculcation weights would be 0.0, and the proportion above would reduce to the variance of cultural norms divided by the average idiosyncratic variance, plus the average measurement-error variance, plus the cultural norm variance. However, the presence of imperfect inculcation reduces the size of this proportion, by making the numerator smaller (since the mean inculcation weight becomes a number between zero and one) and by making the denominator larger (since the variance of inculcation weights becomes a number greater than zero). Thus, the ratio underestimates the proportion of cultural variance in responses to the extent that some respondents are imperfectly inculcated.

5.11 IMPLICATIONS

My rationale for the formal analyses in this chapter was to develop equations for empirical analyses in later chapters. However, this chapter's formal results can also be interpreted in ways that are of interest to survey researchers and to researchers concerned with culture.

5.11.1 Statistical Artifacts

One might not suppose that the variance of individual sentiments about a given concept is contaminated with the mean, because variability is computed around the mean. However, when focusing on a concept with an extreme normative sentiment and when respondents have different levels of cultural inculcation, the variance of sentiments is inflated by the extremity of the norm, as demonstrated in equation (5.6). So variance in sentiments about a concept assesses individual differences in sentiments, and also registers differences in respondents' levels of cultural inculcation and the extremity of the normative sentiment.

This fact has implications for surveys of populations of people if variability of attitudes is interpreted as an index of contention about issues. When surveying a population where people differ in levels of inculcation into the dominant culture, attitudes about one issue may seem more discordant than attitudes about another issue only because the first issue involves a more extreme attitudinal norm than the second.

Additionally, when respondents differ in levels of cultural inculcation, a variable assessing individuals' sentiments around a cultural norm may correlate spuriously with any variable that reflects acquisition of the culture, including measurements of other normative sentiments. This is demonstrated by equation (5.7) and by the equation derived in footnote 2. Consequently, sentiment measurements for normative items, especially items for which the norms are extreme, may aggregate into a general inculcation dimension when factor analyzed. The meaning of such a dimension may be obscure by inspection.

Exactly this phenomenon was found in Thomas and Heise's (1995) factor analyses of ratings data. They analyzed individuals' EPA ratings of multiple concepts and found that the first factor in the correlations among ratings grouped high potency ratings for potent characters such as judge, disciplinarian, and gunman; low potency and activity ratings for low-potency and low-activity characters such as hobo and bum; high-activity ratings for lively characters such as youngster; high evaluation ratings for valued characters such as grandparent, and low evaluation ratings for disparaged characters such as delinquent (Thomas and Heise 1995, p. 430). This factor formed because the least inculcated respondents gave near-neutral ratings on EPA scales, and well-inculcated respondents gave polarized EPA ratings. Thereby the concepts with the most extreme sentiment norms became markers of the general inculcation factor. The signs of factor loadings corresponded to the signs of the norms: for example, judge is potent and hobo is impotent, grandparent is valued and delinquent is disvalued. Therefore, judge potency and grandparent evaluation had positive factor loadings, while hobo potency and delinquent evaluation had negative loadings.

In general, if items assess cultural norms and the sample includes respondents with various levels of cultural inculcation, the main underlying dimension in the correlations may represent inculcation rather than a more substantive latent construct. As mentioned in Chapter 2, Kahneman (1963) applied a similar interpretation to the problem of concept-scale interaction in semantic differential ratings.

5.11.2 Consequences of Enculturation

Equation (5.9) shows that being well inculcated into a culture increases the variance of a person's sentiments about different issues, and thereby inculcation is the functional equivalent of learning sentiments by diverse experiences in different settings. The enculturated sentiments enable intuitive participation in cultural processes (Heise 2007).

Mutual inculcation of a culture by multiple individuals increases the correlation of the individuals' sentiments, as if they shared sentiment-forming experiences in different settings, as shown in equation (5.10). Intersubjectivity is one consequence, in that people with similar sentiments are prone to understand events in similar ways (Heise 2007:14). Moreover, persons with similar sentiments react emotionally in the same way to the actions of an outside figure, providing them with a key prerequisite for a sense of solidarity (Heise 1998).

People who are well indoctrinated into their native culture may have a moderate level of inculcation into a foreign culture because of sentiment correlations across cultures. For example, almost two-thirds of the variance in meanings of emotion terms is shared across American and Japanese cultures (Romney et al. 2000); thus, a Japanese has a fair feeling for emotion interpretations provided by Americans. Additionally, Heise (2001a; 2007, pp. 17–19) revealed remarkably high correlations among the sentiments of Americans, Canadians, Irish, Germans, Japanese, and Chinese with regard to identities and social behaviors. Thus, people from any of these cultures have above-zero inculcation levels with regard to such sentiments in the other cultures. In some cases (e.g., Canada and the United States) the cross-national correlations are so high that inculcation in one culture is nearly the same as inculcation into the other.

5.12 CHAPTER HIGHLIGHTS

- Surveys of culture are prone to coverage errors more than sampling errors. Coverage errors arise when relatively few behavior settings are selected and when respondents within the settings do not include high-level authorities.
- Respondents in a culture survey typically include culture experts but also people who are still being enculturated, perhaps some persons who are poorly socialized, and transients whose sentiments represent cultures other than the one of interest. Thus, some respondents err in their answers to questions about culture because of their low levels of inculcation into the culture of interest.
- Errors that arise in surveys of populations from overly complex questions usually do not arise in surveys of cultural sentiments because stimuli typically are single words or short phrases. However, the terseness of

stimuli means that errors may occur because the desired sense of a stimulus is poorly established through context.

- The contemporary computerized instrument for measuring sentiments moderates several sources of error that arise in surveys. Errors from unfamiliarity with the instrument are obviated by a tutorial. Errors due to incomprehension of a stimulus are largely converted to item nonresponse because respondents are allowed to skip stimuli. Errors due to ennui are combated by various tactics that help respondents remain attentive, and the errors that do occur are distributed evenly across stimuli by randomizing the order of presentation of stimuli.
- A number of kinds of error are similar in surveys of individuals and in surveys of culture. Errors arising from persons refusing to participate occur in both types of surveys. Measurement errors associated with temporal variations and with fitting a subjective answer into an objective response occur in both types of surveys.
- An index of cultural *inculcation*, defined in equation (5.15), corresponds approximately to the regression weight for predicting a respondent's sentiments from cultural norms.
- An index of cultural *commonality*, defined in equation (5.17), corresponds to the correlation between a respondent's sentiment ratings and the average sentiment ratings of others.
- In culture surveys, ideal respondents all give the same response, so there is no meaningful variance in the responses to a single item, and the reliability of the item is zero. However, the commonality index assesses an individual respondent's reliability in reporting norms.
- An overall reliability coefficient for measurements of given concepts by given respondents, defined in equation (5.29), treats variations in average responses to different concepts as the meaningful variance for computing reliability.
- When surveying individuals who differ in levels of inculcation into the dominant culture, sentiments about one issue may seem more diverse than sentiments about another issue only because the first issue involves a more extreme norm than the second. Additionally, individuals' sentiments may correlate spuriously with any variable that reflects acquisition of the culture, including measurements of other sentiments.
- Inculcation into a culture is the functional equivalent of learning a variety of sentiments through diverse experiences. Mutual inculcation of a culture by multiple individuals increases the correlation of the individuals' sentiments, as if they shared sentiment-forming experiences.

6 Correlates of Enculturation

Psychological anthropologists, using quantitative assessments of respondents' levels of enculturation, have determined some characteristics of persons whose answers accord with cultural norms, as opposed to those whose answers have low commonality. Generally speaking, a high-commonality individual displays "the behavioral characteristics of an expert" (D'Andrade 1987, p. 200) in that such a person generally:

- Is bettered educated with regard to issues arising in the normative domain
- Is judged to have more intellectual ability with regard to topics in the normative domain
- Is more experienced in the normative domain and has experienced normative conditions rather than deviant conditions
- Gives the same answers to questions repeated on multiple occasions
- Maintains logical coherence when answers to some questions imply specific answers on other questions

"This pattern of relationships has been found in a wide variety of cultural domains, including color, beliefs about disease, plant taxonomies, and American parental sanctions for rule breaking" (D'Andrade 1995, p. 213). Additionally, Rossi and Berk (1985, p. 341) report that more formal education is also associated with high commonality in rating the seriousness of crimes and in rating the prestige of occupations.

In this chapter we examine correlates of respondents' high versus low commonality in sentiment ratings. The characteristics of high-commonality respondents in anthropology may not generalize to respondents with high commonality in sentiments because the anthropological studies have focused on cognitive knowledge rather than on attitudes and sentiments. In the next section I describe the indices of commonality and cultural inculcation that I computed and examine some statistical characteristics of the indices. Then I examine how respondents' behavior as raters related to their inculcation and commonality measures. I use the results of the analyses of rater conduct to purge the sample of a small number of uncooperative respondents. After recomputing

Surveying Cultures: Discovering Shared Conceptions and Sentiments, By David R. Heise

the inculcation and commonality indices in the purified sample of respondents, I scan for correlates of the indices in a series of sections focusing on demographics, academic activities, and nonacademic activities.

6.1 INDICES

The empirical study described in Chapter 3 presented each respondent with 100 concepts (stimuli) for rating on the Evaluation, Potency, and Activity (EPA) dimensions. The overall study included 15 sets of 100 concepts, with each respondent assigned randomly to one of the sets. I regressed each respondent's ratings on the mean ratings of all other respondents[1] who rated the same 100 concepts, separately for Evaluation, Potency, and Activity. According to equation (5.15), the regression coefficients from these analyses index respondents' levels of cultural inculcation. I also computed the correlation of each respondent's EPA ratings of the 100 concepts with the mean ratings from other respondents. According to equation (5.18), these correlation coefficients index respondents' commonalities in cultural sentiments, taking into account their cultural inculcations, personal variations in sentiments, and measurement errors.

Computed over all the respondents in this study, the Evaluation and Potency inculcation indices share 64 percent of their variance, the Evaluation and Activity inculcation indices share 66 percent of their variance, and the Potency and Activity inculcation indices share 60 percent of their variance. The Evaluation and Potency commonality indices (i.e., correlation coefficients) share 66 percent of their variance, the Evaluation and Activity commonality indices share 67 percent of their variance, and the Potency and Activity commonality indices share 59 percent of their variance.

The commonality indices are related to the inculcation indices by a quadratic function, and when incorporating this function in the computation of associations, the inculcation indices account for 84 percent of the variance in the commonality indices on the Evaluation dimension, 78 percent on the Potency dimension, and 74 percent on the Activity dimension.

The distributions of the commonality indices were skewed toward high values, except for small bulges at very low levels, with means of 0.75, 0.54, 0.54 and modes of 0.89, 0.64, 0.72 for Evaluation, Potency, and Activity, respectively. The distributions of the Evaluation, Potency, and Activity inculcation indices were bell-shaped around means close to 1.0. Again, the lower tails bulged.

[1]The equations derived in Chapter 5 define expectations in indefinitely large samples. Since the number of raters for each stimulus ranged from 64 to 88, I related each respondent's rating to the mean rating computed without that respondent's rating in order to eliminate artifactual inflation of the covariance.

Analyses in this chapter are conducted with respondents from all 15 stimuli sets pooled together.[2] This gives a total N of 1,113 for most variables (400 Arts and Sciences students—270 Midwestern, 130 Eastern—and 713 Business School students). However, some variables were measured only at time 1 or time 2 in the Business School survey, and in those cases the sample size is 642 or 713 rather than 1,113.

6.2 CONDUCT AS A RATER

I split the distributions of the inculcation and commonality indices into two groups for tabular analyses. Respondents in the bottom quintile represent those with low commonality and low inculcation; respondents in the upper four quintiles represent those with acceptably high commonality and inculcation. The breakpoints were 0.65, 0.40, 0.38 for EPA commonality indices, and 0.76, 0.63, 0.64 for EPA inculcation indices. Table 6.1 shows the relations between commonality and several aspects of respondents' conduct during the rating task.

TABLE 6.1 Relations Between Individual–Norm Correlation and Characteristics of Respondents' Rating Performance

	Chi-Square	Degrees of Freedom	Significance
	Minutes Worked (20.9 or Less; 21–26.9; 27–35.9; 36 or More)		
Evaluation	126.783	3	0.000
Potency	102.474	3	0.000
Activity	138.650	3	0.000
	Number Skipped (none; 1–3; 4–7; 8–20)		
Evaluation	110.389	3	0.000
Potency	95.112	3	0.000
Activity	118.893	3	0.000
	Polarization (2 or Less; 2.1–6.9; 7 or More)		
Evaluation	43.939	2	0.000
Potency	31.377	2	0.000
Activity	41.213	2	0.000

Note. N = 1113 for all analyses. The second dimension of each cross-tabulation consists of correlation coefficients between respondents' ratings and mean ratings, split at first versus higher quintiles.

[2] Pooling respondents assumes that the average inculcation weight is about the same within all sets of raters. The assumption is plausible since respondents for each stimuli set were drawn randomly from the same body of raters. Analyses of variance show that differences in the means of the inculcation indices are insignificant ($p > 0.5$) across sets of raters on all three dimensions.

The amount of time that a respondent devoted to the task was predictive of the respondent's commonality on all three dimensions. Respondents who took less than 21 minutes to answer the demographic questions, take the tutorial, and rate 100 concepts on three dimensions were less likely to be in the high-commonality group: 54 percent, 56 percent, and 52 percent for Evaluation, Potency, and Activity, respectively. Eighty-four percent or more of respondents who took 21 minutes or more had high-commonality indices.

Another predictor of commonality on all three dimensions was number of stimuli skipped. (Respondents were supposed to skip a stimulus if they did not know the meaning of the concept.) Perhaps surprisingly, respondents who skipped few were least likely to have high commonality. Of those who skipped none, 62 percent had high commonalities on Evaluation, 64 percent on Potency, and 62 percent on Activity. Of those who skipped one to three stimuli, 79 percent had high commonalities on Evaluation, 78 percent on Potency, and 77 percent on Activity. Among those who skipped four to 20 stimuli, 91 percent or more were in the high commonality group on all three dimensions.

Polarization is the mean of the squares of a respondent's ratings, averaged across all three EPA dimensions and across all stimuli that the respondent rated. Polarization is near zero if a respondent barely flicked the pointer off center for each rating (the rating program requires some movement of the pointer before the next stimulus–scale combination is shown). Polarization could be above 18 if most of a respondent's ratings were made at the endpoints of the scales. Polarization of 2 or less corresponded with a low likelihood of having high commonality on all three dimensions: 48, 55, and 49 percent for E, P, and A. This compares with percentages above 80 for respondents with polarization above 2, on all three dimensions.

I experimented to find breakpoints for the performance variables other than skipped stimuli (a variable discussed further in Section 6.4), in order to discriminate respondents with near-zero commonalities from other respondents.

- *Rushed ratings.* One of the 29 respondents who devoted 13 minutes or less to the task had high-commonality indices on Evaluation and Activity, two had high-commonality indices on Potency. Meanwhile, 82 percent of those devoting 14 minutes or more had high-commonality indices on all three dimensions.
- *Near-zero ratings.* None of the 17 respondents with a polarization of less than 1.0 had high-commonality indices on Evaluation and Activity, one had high-commonality indices on Potency. Meanwhile, 81 percent of those with polarizations of 1.0 or more had high-commonality indices on all three dimensions.
- *Combined criteria.* Of the 41 respondents who either rushed or who had polarization of 1.0 or less, one (2 percent) had high commonality on

Evaluation and Activity, and three (7 percent) had high commonality on Potency. Of the remaining respondents, 81 percent had high-commonality indices on all three dimensions.

The combined criteria identify low-commonality respondents almost exclusively. That is, nearly all respondents who went through the ratings very fast or who made most ratings close to the midpoint of the scales had low commonality.

Negative Commonality. Seven respondents gave ratings that had significant negative correlations (–0.17 or less) with sentiment norms on Evaluation, Potency, or Activity. Apparently, these respondents were either reporting the opposites of their own sentiments, which conformed to norms, or were members of a reactive culture in which sentiment norms tend to be opposites of standard norms (Cohen 1955). Since reactive cultures have been discredited for their original purpose of explaining delinquency (Gordon et al. 1963), and no empirical evidence of reactive cultures has appeared, purposeful falsification seems the most likely interpretation of the negative correlations.

6.3 PREDICTING CULTURAL AUTHORITATIVENESS

The analyses of rater conduct suggested that a small proportion of respondents participated insincerely in the study, providing data that did not authentically represent their personal sentiments. Specifically, 44 respondents (4 percent) displayed one or another of three criteria of insincerity: rushing (i.e., devoting 13 minutes or less to the survey), near-zero ratings (i.e., polarization of 1.0 or less), or significantly negative commonality (i.e., commonality indices less than –0.17 on Evaluation, Potency, or Activity).

To see if their data affected results, I computed one-way analyses of variance of the enculturation indices with all the variables in the following sections, when including the insincere respondents, and again with them excluded. More than a dozen effects changed in significance at the 0.05 level between the purified and unpurified samples. Overall, the results of analyses conducted with insincere respondents included were more complex and difficult to interpret. Therefore, in the following sections I report results of analyses after removing the 44 respondents who displayed the three criteria of insincerity.

I recomputed the commonality and inculcation indices using sentiment norms derived within the purified sample. Bulges in the lower tails of the distributions of the indices were eliminated with the insincere respondents removed. The ranges of inculcations on Evaluation, Potency, and Activity were –0.09 to 1.99, –0.40 to 2.35, and –0.35 to 2.30 on Evaluation, Potency, and Activity, respectively. The ranges of commonalities were –0.09 to 1.00, –0.16 to 0.89, and –0.12 to 0.92.

In the following three sections I examine associations of the enculturation indices with demographic, academic, and organizational variables. I employ multiple regression analyses for this purpose, and I include categorical variables in the regression equations—both their main effects and their interactions, as in analyses of variance.

Approximately normal distributions of dependent variables are desirable in statistical analyses to obtain valid significance tests. The distributions of inculcation indices are bell-shaped and reasonable approximations to normality. Commonality indices in their original form are correlation coefficients, and their distributions are skewed to the right. Therefore, I normalized the distributions of the commonality indices with Fisher's r-to-z transformation, as indicated in equation (5.20). The distributions of the transformed commonality indices are bell-shaped.

In preparation for the multivariate analyses, I reconnoitered associations with one-way analyses of variance to determine which variables had significant relations with enculturation variables. These preliminary analyses indicated that all available variables except part-time work experience needed to be included in the multiple regressions.

Some categorical variables were recoded for clarity or as a consequence of posthoc analyses conducted in conjunction with the one-way analyses of variance.

- Gender was turned into the dummy variable, Female = 1, not = 0.
- Age was converted into the dummy variable, Older = 1, not = 0, with Older being 21 years or more.
- Race was converted into the dummy variable, Asian- or African-American = 1, not = 0. Post hoc analyses indicated that Asian- and African-Americans were similar in their enculturation scores.
- Geographic origin was converted into the dummy variable, Northeast = 1, not = 0.
- Respondent's school was turned into the dummy variable, Business School = 1, not = 0. The zero category consists of Arts and Sciences students at both Indiana University and the University of Connecticut. The Business School students were all at Indiana University.
- The marriage variable was converted into the dummy variable, Ever married = 1, not = 0.
- Full-time work experience was turned into the dummy variable, Full-time worker = 1, not = 0. Respondents were included in the full-time worker category only if they had more than two years of full-time work experience.

Age, the academic variables, and the organizational variables were assessed only among the Business School respondents. Therefore, I present the results

TABLE 6.2 Regression Coefficients for Predicting Inculcation and Commonality from Personal Characteristics

	Inculcation			Commonality		
	E	P	A	E	P	A
Ever Married	−0.041	−0.075	−0.209	−0.052	−0.059	*−0.104*
Female	0.105	0.007	0.060	*0.149*	0.027	0.068
Asian-, African-American (AAA)	*−0.211*	−0.197	*−0.281*	−0.180	−0.108	*−0.161*
Northeast (NE)	0.000	0.008	0.011	0.052	0.052	0.054
Business School	−0.097	−0.054	−0.115	*−0.088*	−0.018	*−0.058*
Female × AAA	0.094	0.140	0.052	−0.002	0.029	−0.017
Female × NE	0.060	0.077	0.059	0.022	−0.020	−0.015
Female × Business School	0.112	*0.128*	0.133	0.127	0.057	*0.065*
AAA × NE	0.171	0.196	−0.072	0.068	0.131	−0.099
AAA × Business School	0.081	0.085	0.080	−0.046	−0.042	−0.014
NE × Business School	−0.117	−0.009	−0.145	−0.132	−0.020	−0.131
Female × AAA × NE	−0.235	−0.093	0.078	−0.080	−0.033	0.118
Female × AAA × Business School	−0.181	−0.296	−0.244	−0.132	−0.080	−0.050
Female × NE × Business School	0.083	−0.018	0.278	0.101	0.057	0.292
AAA × NE × Business School	−0.544	*−0.993*	−0.214	−0.484	*−0.616*	−0.104
R^2	0.113	0.037	0.094	0.132	0.053	0.118

Note. Italic type signals statistical significance with $p \leq 0.10$, an underline signals $p \leq 0.05$, both signals $p \leq 0.01$. Few respondents were Ever Married, so the variable is treated as a covariate. The four-way interaction is excluded because of few cases in some cells.

of two sets of regression analyses. Table 6.2 covers the full sample, and Table 6.3 covers just the respondents from the Business School. Both tables show how various predictors relate to inculcation and commonality on the Evaluation, Potency, and Activity dimensions.

In Table 6.2 the variable "Ever Married" is included just as a covariate rather than as a factor combining with other factors because only 23 respondents were in the ever-married category. "Female," "Asian- or African-American," "Northeast," and "Business School" are treated as factors in an analysis of variance design, and their two-way and three-way interactions are included in the regressions. The four-way interaction involving all factors was excluded because no respondents were in the sample to assess the effect of the combination.

In Table 6.3 "Grade-Point Average" "Learning Is Fun," "Quit Risky Class," "Cut Class in Easy Course," "Browse the Library," and "Group Participation" are treated as covariates. For the regression analyses these variables' multiple categories were coded on an assumed-interval scale, and the scales had linear relations to enculturation indices, when they had any association at all. Factors in the analysis of variance design are "Female," "Asian- or African-American," "Older," and "Full-Time Worker."

TABLE 6.3 Regression Coefficients for Predicting Inculcation and Commonality from Personal Characteristics: Business School Respondents Only

	Inculcation			Commonality		
	E	P	A	E	P	A
Grade-Point Average	0.008	0.004	0.004	0.050	0.020	0.026
Learning Is Fun	*0.056*	*0.063*	0.048	0.080	0.056	0.043
Quit Risky Class	−0.009	−0.008	−0.024	−0.030	−0.016	*−0.023*
Cut Class in Easy Course	−0.014	−0.003	0.019	*−0.033*	−0.011	0.002
Browse the Library	−0.024	−0.017	−0.026	*−0.051*	−0.028	−0.028
Group Participation This Semester	*0.036*	0.032	*0.045*	0.020	0.010	*0.016*
Female	*0.237*	*0.193*	*0.204*	*0.250*	*0.096*	*0.125*
Asian-, African-American (AAA)	−0.049	−0.096	−0.113	−0.208	−0.177	−0.142
Older	0.038	0.056	−0.009	−0.018	−0.002	−0.040
Full-Time Worker	*−0.374*	*−0.342*	*−0.454*	*−0.445*	*−0.240*	*−0.263*
Female × AAA	−0.047	−0.121	−0.262	−0.063	−0.020	−0.129
Female × Older	−0.085	−0.170	−0.084	−0.010	−0.054	−0.029
Female × Full-Time Worker	*0.333*	0.342	*0.579*	*0.323*	*0.249*	*0.350*
AAA × Older	−0.060	0.039	−0.041	0.062	0.102	−0.008
AAA × Full-Time Worker	*0.595*	0.771	*0.889*	*0.674*	0.418	*0.474*
Older × Full-Time Worker	*0.357*	*0.453*	*0.472*	*0.365*	*0.314*	*0.272*
Female × AAA × Older	−0.153	−0.131	0.058	−0.118	−0.050	0.210
Female × Older × Full-Time Worker	−0.312	−0.444	−0.614	*−0.450*	−0.427	−0.393
AAA × Older × Full-Time Worker	*−0.940*	*−1.333*	*−1.542*	*−0.950*	*−0.746*	*−0.780*
R^2	0.171	0.081	0.156	0.223	0.120	0.192

Note. Italic type signals statistical significance with $p \leq 0.10$, an underline signals $p \leq 0.05$, both signals $p \leq 0.01$. The four-way interaction and the following interaction were excluded because of empty cells, or collinearity with included variables: Female × AAA × Full-Time Worker.

("Northeast" was not included because only nine business school students were from the Northeast.) Two- and three-way interactions of the factors are included in the regressions. However, the four- and five-way interactions plus some of the two-way and three-way interactions are excluded because of empty cells in the cross-tabulation of factors or because the interaction effects are collinear with lower-order effects.

Both tables indicate three levels of statistical significance: $p \leq 0.10$, $p \leq 0.05$, and $p \leq 0.01$. I will refer to effects significant at the 0.10 level as "marginally significant," and those significant at the 0.01 level as "highly significant."

The bottom rows of the tables show the proportions of variance explained by the various predictors (R^2). Although an equation constant is significant in every regression equation, I do not show the constants in the tables,[3] and I treat them all as zero in the following exposition: a linear transformation of no consequence in presenting the results.

6.3.1 Demographics

Consider first the full sample (Table 6.2). Females were significantly more normative than males with respect to some indices. A female who was not Asian- or African-American, and was not from the Northeast, was predicted to have an Evaluation inculcation of 0.105, an Evaluation commonality of 0.149, and an Activity commonality of 0.068. (The predictions are for respondents who were not in Business School; the Business School factor will be considered under academic predictors.) A corresponding male would have had scores of zero. The gender differences also apply to individuals who were Asian- or African-American and from the Northeast, although with slightly different values if one uses nonsignificant regression coefficients when computing the predicted values.

With marginal significance, Asian- and African-Americans were lower in some types of enculturation than other ethnicities. The predicted Evaluation inculcation was –0.211 lower, the predicted Potency inculcation was –0.281 lower, and the predicted Activity commonality was –0.161 lower. Potency inculcation and commonality were significantly lower for the one respondent who was a Northeast Asian- or African-American in Business School.

Respondents who had been married were lower in Activity enculturation than were the unmarried, the effect being significant in the case of inculcation and marginally significant in the case of commonality. However, relatively few of the respondents were married or previously married, 27 in all. In the full sample, no other demographic factors differentiated respondents significantly with regard to enculturation on Evaluation, Potency, or Activity.

Now consider the demographic variables in just the subsample of Business School respondents (Table 6.3). Again, females had higher predicted enculturation than males, and the regression coefficients for both indices were highly significant on all three dimensions. There was one significant complication in predicted Potency inculcation. An older female was predicted to have a Potency inculcation score of (0.193 – 0.170) = 0.023; the predicted score for a corresponding older male was zero, so a gender difference was lacking in this particular instance.

In the Business School subsample, Asian- or African-Americans were significantly lower than other ethnicities in commonalities on all three dimensions. No other demographic factors were significant in the Business School subsample.

[3]In the full sample the constants were—Inculcation E: 1.027, P: 1.008, A: 1.037; Commonality E: 1.113, P: 0.661, A: 0.662. In the Business School subsample the constants were—Inculcation E: 0.769, P: 0.847, A: 0.798; Commonality E: 0.700, P: 0.506, A: 0.455.

6.3.2 Academic Variables

Table 6.2 provides information regarding the relations between enculturation and a respondent's general area of study—Arts and Sciences versus Business School. The table reveals that males in Business School had lower enculturation than other respondents, at significant or marginally significant levels. Specifically:

- The predicted Evaluation inculcation of males in Business School was –0.097; for females in the Business School the prediction was (–0.097 + 0.112) = 0.015, and for other respondents the prediction was zero. For Evaluation commonality, the predictions are –0.088 for Business School males, 0.039 for Business School females, and zero for everyone else.
- The predicted Potency inculcation of Business School males was –0.197; for females in the Business School the prediction was (–0.197 + 0.128) = –0.069. The prediction for Arts and Sciences respondents was zero.
- The predicted Activity inculcation of males in Business School was –0.115; for females in the Business School the prediction was (–0.115 + 0.133) = 0.018, and for other respondents the prediction was zero. For Activity commonality, the predictions were –0.058 for Business School males, 0.007 for Business School females, and zero for Arts and Sciences respondents.

Table 6.3 shows the relations of enculturation to other academic variables. Grade-point average was not significantly related to inculcation, but students with higher grades tended to be have higher Evaluation and Activity commonality. In particular, respondents with a GPA of less than 2.0 had predicted values of zero, whereas those four intervals higher with a GPA of 3.6 or more had a predicted Evaluation commonality of 0.200 and a predicted Activity commonality of 0.104.

At marginal significance, the predicted Evaluation inculcation of respondents who agreed that learning is fun was 0.112 higher than the zero value predicted for those who disagreed (the difference was twice the value of the regression coefficient since two intervals separate agreeing from disagreeing). The predicted difference on Potency inculcation of 0.126 was also marginally significant. Predicted commonality differences of 0.160 on Evaluation, of 0.112 on Potency, and of 0.086 on Activity all were significant.

Anticipated actions regarding courses that are very hard or very easy had no significant relations with enculturation. (However, one relation to a commonality index was marginally significant in each case.) Students who reported browsing the library had significantly lower commonality on all three dimensions. This seems contrary to the results obtained for grade-point average and attitude toward learning, if browsing the library is interpreted as being an indicator of serious scholarship. Perhaps students who read widely were more likely to develop individualistic sentiment norms; or, viewing the causality in the other direction, perhaps some students with low cultural inculcation escaped to the library to get away from frustrating cultural experiences.

6.3.3 Organizational Activity

Business School respondents with more than two years' experience as a full-time worker were significantly lower in enculturation indices than other Business School respondents (see Table 6.3). However, the factor interacted significantly with age and ethnicity. (In the following I forgo showing the composition of each prediction in terms of separate regression coefficients.)

- Predictions for an experienced worker who was young and not Asian- or African-American (23 cases). *Inculcation*—E: –0.374; P: –0.342; A: –0.454. *Commonality*—E: –0.445; P: –0.240; A: –0.263.
- Predictions for an experienced worker who was older and not Asian- or African-American (27 cases). *Inculcation*—E: 0.021; P: 0.167; A: 0.009. *Commonality*—E: –0.098; P: 0.072; A: –0.031.
- Predictions for an experienced worker who was young and Asian- or African-American (four cases). *Inculcation*—E: 0.173; P: 0.333; A: 0.332. *Commonality*—E: 0.021; P: 0.001; A: 0.069.
- Predictions for an experienced worker who was older and Asian- or African-American (four cases). *Inculcation*—E: –0.433: P: –0.452; A: –0.798. *Commonality*—E: –0.520; P: –0.331; A: –0.487.

The value predicted on all indices was 0.000 for young respondents who were not Asian- or African-American and who had two years' experience or less as a full-time worker. The breakdown above reveals that most of the cases of experienced workers with lower values on enculturation indices were young respondents other than Asian- or African-Americans (in fact, all were white and mostly males), with a few additional cases being older Asian- or African-Americans. Meanwhile older respondents who were not Asian- or African-Americans and young Asian- or African-Americans were about average on most enculturation indices and in some instances even were notably higher than average.

Greater group participation was associated with more sentiment inculcation. Someone who participated in no extracurricular groups in the semester of the survey had a predicted value of zero on the inculcation indices. Someone who participated in five or more groups was four intervals higher with predicted inculcations of 0.144 on Evaluation, 0.128 on Potency, and 0.180 on Activity. The commonality on Activity predicted was also higher at marginal significance: 0.064 for a respondent in five or more groups. This supports the findings of Thomas and Heise (1995, p. 434). They found that high-commonality respondents were linked socially into organizations and family and that they reported having close friends, whereas low-commonality respondents were social isolates who were uninvolved with organizations, uncommitted to family, and who did not have close friends.

6.4 IMPLICATIONS

The initial analyses reported in this chapter considered some characteristics of respondents' performance as they were completing the survey: minutes worked, average distance that they moved the pointer on the scale, and number of stimuli that they skipped. Significantly negative correlations between a respondent's ratings and the mean ratings of everyone else constituted another indirect indicator of respondent performance. The negative correlations could indicate that the respondent belongs to a reactive culture in which sentiment norms are contrary to standard norms, or that the respondent participated inauthentically in the survey, often rating stimuli in opposition to their own sentiments. Whichever the case, respondents whose ratings correlate quite negatively with the ratings of others do not provide useful contributions regarding sentiment norms.

Explorations of relations between these variables and adequacy as an informant concerning sentiment norms produced some guidelines for identifying respondents who participate insincerely in Internet surveys of sentiments. Some respondents gave adequate assessments of sentiment norms even though they worked very fast. Beyond a certain speed, however, correlation between the respondent's ratings and the ratings of others vanished, suggesting that respondents who worked at such speeds were simply going through the motions of rating rather than sincerely considering the stimuli. Almost no respondents who spent less than 13 minutes on the entire survey had ratings that correlated appreciably with the mean ratings of other respondents. The 13-minute cutoff translates to a more general measure of about 2½ seconds per screen presented, given that the survey presented 315 screens, including background questions and tutorial as well as sentiment ratings for 100 stimuli on three dimensions. Respondents who spend less than 2½ seconds per screen almost certainly are not providing useful contributions.

The instrument used for measuring sentiments requires moving the pointer away from the center of the scale to some point between the center and an endpoint. The center is coded zero, the endpoints are coded plus or minus 4.3, and the median distance of each move by all respondents in this study was 2.2 units. However, 1.5 percent of the respondents moved the pointer less than one unit on average, and the ratings of these respondents mostly did not predict ratings of other respondents. Thus, it appears that respondents who barely move the pointer either are not inculcated with sentiment norms or are not reporting their sentiments sincerely.

Some respondents who skipped no stimuli provided ratings that correlated poorly with the ratings of respondents who skipped one or more stimuli (the median number skipped was three). The measurement software used in this project penalized skipping with a confirmation dialogue followed by a 5-second delay before the respondent was allowed to continue the rating task. Respondents who were participating insincerely would have discovered during the tutorial that skipping slowed them, and evidently they avoided the option. On the basis of these results, I changed the rating program, removing the delay after the respondent confirms a skip. The rationale is that in future projects it is better to have insincere respondents skip stimuli prodigally, providing no data, than to have them enter erroneous ratings.

An unusual proportion of respondents were unfamiliar with the concept of "racketeer," so the inclusion of this stimulus in the test–retest survey fortuitously provides a basis for examining how respondents dealt with unfamiliar concepts. Thirteen of the 42 respondents who were presented with the stimulus at time 1 used the skip function at either time 1 or time 2. However, just one person skipped the stimulus both times. Among the other 12, half skipped the first time and half skipped the second time. Nine respondents skipped the stimulus at one time but rated it as neutral or positive at the other time. Three respondents skipped the stimulus at one time and offered a normative EPA profile at the other time. From these patterns, one may deduce that respondents confronted with an unfamiliar word sometimes used the skip function, and at other times they rated the stimulus as neutral, or some random value. However, the nonpunishing implementation of the skip function may increase its appropriate usage.

6.4.1 Culling Respondents

Although incongruous with the individualistic focus of survey research, culling a sample of respondents to eliminate those who provide worthless data is consistent with the normative focus of ethnological research, merely turning around the standard ethnographic practice of seeking out informed and cooperative culture informants.

Consider first a significantly negative correlation between a respondent's ratings and the mean ratings of everyone else (i.e., a negative inculcation or commonality score), which indicates that a respondent is expressing a reactionary culture or else is trying to subvert the research study by reporting the opposite of what the respondent actually believes. In either case the respondent should be removed from the sample of respondents if the sole function of the study is to assess cultural norms.

An argument also can be made for culling respondents with low-commonality scores—say, less than 0.20, paralleling the traditional argument in test construction of culling items with low item-to-total correlations. As

Nunnally (1967, p. 242) said, such items, or respondents, "should be carefully inspected" to determine if they add enough valid information to be worth including.

Respondents who spent 13 minutes or less for the survey considered here devoted no more than 2.5 seconds on average to each presented screen. Such a high degree of rushing yielded ratings that were uncorrelated with other respondents' ratings. Thus, respondents who race through the task reasonably may be culled since their ratings offer little valid information about sentiment norms. However, the time criterion must be applied with restraint, since a sizable number of respondents in this study provided excellent data while giving only 20 minutes to the task: that is, about 3.8 seconds per screen. Respondents who frequently moved the pointer barely past the "slightly" marks on the scale produced ratings correlating poorly with ratings of other respondents. Accordingly, it is reasonable to cull respondents with an average polarization of 1.0 or less.

The sentiment-rating program instructs respondents to skip concepts that they do not understand. In this study, respondents who skipped no stimuli at all provided ratings that correlated poorly with ratings of other respondents. Thus, the skip function was not being used by these respondents as often as they should have been using it. That undoubtedly was because I had built in a 5-second punishing wait to discourage overuse of the function, and the punishment ended up being too effective. Accordingly, I now have changed the sentiment-rating program such that respondents using the skip function still must affirm their decision to skip by clicking a second button on a pop-up window, but they no longer have to suffer a 5-second delay. With this change in the program, there probably will be few respondents who skip no stimuli at all, so dropping such respondents will not be an efficient method of culling. On the other hand, there probably will be more respondents who can be dropped for displaying lack of commitment to the task by skipping high percentages of stimuli.

6.4.2 Individual Characteristics

In the second part of this chapter I considered which characteristics of respondents predict the normativeness of their sentiments. These results have limited generalizability and limited potential for causal interpretation since they are based on a sample designed for identifying cultural norms rather than on a probability sample that represents a population and defines variations among individuals in the population.

The analyses indicated that females were more normative than males. This difference was most pronounced in the Business School subsample, suggesting that the difference was influenced by selection processes. For example, the males who entered the school of Arts and Sciences may have had more normative sentiment norms than males in general, reducing the gender difference in the overall sample. Alternatively, the males who entered the Business

School may have had sentiments that departed from norms to an unusual degree. Investigations of middle-class cultures in more diverse locales than universities are needed to see if young white females are high inculcators because their own particular culture is being tapped or because such respondents simply are well socialized into the general middle-class culture.

Older females in the Business School had potency sentiments that departed from norms—similar to males'—which perhaps is a result of anticipatory socialization of some of the females into the business world.[4] Simpson and Stroh (2004, p. 720) found that women who adopted a masculine pattern of suppressing positive emotions and simulating negative ones were least likely to report feelings of inauthenticity at work, and they speculated that the masculine pattern of affective display "may somehow be linked to the same attributes (e.g., assertiveness) that account for these women's success in male-typed jobs."

In the Business School, Asian- and African-Americans were less normative in their sentiments than other ethnicities. Some similar differences existed in the overall sample but were not significant or were only marginally significant. Consequently, selection processes, similar to those proposed for gender, might have been at work, thereby accounting for the differences between the overall sample and subsample.

In the Business School, serious students were more normative in their sentiments than students less committed to learning. A high grade-point average predicted greater normativeness with respect to some indices. Expressing a positive attitude toward learning was even more related to adoption of sentiment norms, although that might be simply because such an attitude is itself a normative sentiment within the university environment.

Among Business School students, more than two years' experience as a full-time worker was associated with less adoption of sentiment norms, especially among young whites. Also among Business School respondents, participation in a greater number of social groups was significantly related to normative inculcation.

Even combined, these factors provided only modest power in predicting variations in individuals' degree of normativeness in sentiments. In the overall sample, all factors combined accounted for about 10 percent of the variance in indices of Evaluation and Activity normativeness, and around 5 percent in the case of Potency (see the R^2 values in Table 6.2). The greater number of predictors available for Business School respondents led to higher levels of explained variance, but the numbers still range just from 16 to 22 percent in the case of Evaluation and Activity and from 8 to 12 percent in the case of Potency (see the R^2 values in Table 6.3). Thus, the preponderance of differences among individuals in their degree of normativeness is unexplained.

On the other hand, the sentiment ratings of most respondents do predict norms reasonably well. The median correlations of a respondent's ratings with

[4]I am grateful to Professor Claire Francis for this idea and for the associated reference.

the mean ratings of others were 0.83 for Evaluation, 0.60 for Potency, and 0.59 for Activity (in the unpurified overall sample). Thus, the range is mostly from adequate informants regarding culture to superb informants.

Attributes of better respondents in assessing sentiment norms bear some similarity to the attributes of expert informants regarding other aspects of culture (D'Andrade 1987, p. 200). In both cases, more authoritative persons tend to have more intellectual achievement and zeal. Better respondents in assessing sentiment norms tend to be female, which may relate to being better educated and more experienced in the domain of affect: "A growing body of research suggests girls are socialized to be more attuned to emotions than are boys" (Cross and Madson 1997, p. 14). The education–experience attribute might also relate to the relatively lower expertise of Asian- and African-Americans: a recent study showed that at least some African-Americans are inculcated into a culture different than white middle-class culture (Sewell and Heise 2009).

6.5 CHAPTER HIGHLIGHTS

- Respondents who spent less than 2.5 seconds per screen or who flicked the pointer of the rating scale barely off neutral probably were not reporting their sentiments, since the commonalities of almost all of these respondents were very low.
- Ratings of a few respondents had significant negative correlations with the ratings of others. These respondents might have been falsifying their sentiments purposefully.
- Skipping a stimulus was penalized in this study by a 5-second wait. This was an excessive penalty, since unmotivated respondents avoided using the skip function, whereas other respondents sometimes preferred to guess at sentiment ratings rather than use the skip function.
- On the whole, females had more normative sentiments than males, possibly because females are more attuned than males to affective matters. The sentiments of males in Business School were particularly low in commonality.
- Asian- and African-Americans had somewhat lower commonalities than other ethnicities, possibly because of early socialization into variant cultures.
- Some academic variables related positively to commonality of sentiments, one academic variable related negatively, and several others were unrelated. Thus, academic station related only marginally to sentiment enculturation.
- Sentiments of students with little work experience had higher commonality than the sentiments of students with much work experience. Sentiments of students with many group affiliations were more normative than the sentiments of students with few affiliations.

- Measurements of cultural sentiments can be improved by culling respondents on the basis of several objective measures revealing respondent noncooperation or cultural incompetence.
- The demographic, academic, and organizational variables considered in this study left unexplained more than three-quarters of the variance in respondents' enculturation. However, most respondents were well enculturated, ranging from adequate to superb as informants about sentiment norms.

7 Consensus in Sentiments

Romney, Weller, and Batchelder (1986, pp. 317–318) listed three enabling assumptions for culture-as-consensus methodology in their introductory article on the topic. Rephrasing somewhat, it is assumed that respondents homogeneously represent a common culture; second, that no consideration other than their common culture explains the similarity in respondents' answers to questions about cultural norms; and third, that each respondent has a uniform level of inculcation regarding the norms being considered in a study, although levels of inculcation may differ across respondents.

Romney, Weller, and Batchelder (1986, p. 323) proposed that the adequacy of the first two assumptions can be judged empirically. In brief, the procedure is to compute the correlations between respondents, with the correlations being computed across respondents' answers to different questionnaire items. For example, if 20 respondents rated the goodness of 100 concepts, the correlation matrix would be 20 × 20, and each correlation would be computed over 100 observations. The assumption of homogeneous influence from a single common culture and the assumption that other influences are not generating shared answers imply that the correlation matrix has a definitive factor structure. A single large factor should account for the correlations between respondents, and factors after the first should decline gradually and regularly in size in the manner that is typical when factoring correlations between sets of random numbers. A later publication (Romney, Batchelder, and Weller 1987; p. 173) stated the criterion more specifically while discussing how to compute solutions with a particular statistical package: "The eigenvalues of the unaltered correlation matrix are printed out and if the data fit the model, the first eigenvalue should be a few times larger than the second eigenvalue."

As mentioned, the relevant correlation matrix consists not of the usual correlations of variables computed across respondents. Rather, the focus is on factoring the matrix of correlations among respondents computed across comparably scaled items, called Q-factoring in the factor analysis literature (e.g., Rummel 1970). A determination that the first eigenvalue is much larger than subsequent eigenvalues of gradually diminishing size is a variation of a frequently used judgmental procedure in factor analysis—a scree test—ordinarily

Surveying Cultures: Discovering Shared Conceptions and Sentiments, By David R. Heise

used to fix the number of consequential factors contributing to a matrix (e.g., see Van de Geer 1971, p. 147). As adapted in culture-as-consensus research, the test ascertains that only one factor is consequential.

In a study of college students answering a general information test, Romney, Weller, and Batchelder (1986, p. 323) found a ratio of first to second eigenvalues of 5.15. In a study of college students ranking the importance of causes of fatalities in the United States, the first eigenvalue was about three times the size of the second (Romney, Batchelder, and Weller 1987, p. 174). Romney (1999; p. S108) reported a first-to-second eigenvalue ratio of 13.62 for data on Guatemalan women's judgments about diseases. In all of these cases, the conclusion was that the data fit the assumptions of the model. On the other hand, in a study of college males judging the effectiveness of birth control methods, the first-to-second eigenvalue ratio was 1.18, from which it was inferred that the students did not share knowledge of the domain (1987, p. 175).

In this chapter I assess the adequacy of the consensus assumption in studies of cultural sentiments. As detailed in Chapter 3, each respondent was presented with 100 concepts for rating on the Evaluation, Potency, and Activity dimensions, being assigned randomly to one of 15 different sets of concepts. Within each of the 15 stimuli sets, I computed the correlations of respondents' ratings across 100 stimuli, separately for Evaluation, Potency, and Activity. Ratings that were missing because a respondent skipped a stimulus were filled with the respondent's mean rating on other stimuli for the corresponding dimension before computing correlations. Altogether 1,113 respondents were involved in these analyses: 400 Arts and Sciences students and 713 Business School students. I then conducted principal component analyses of each of the 45 correlation matrices (15 stimuli sets × three dimensions) and examined eigenvalues in the manner suggested by Romney and his colleagues in order to assay how well the data fit the assumptions of the culture-as-consensus model.

The sections following report the results of this work and additionally note some continuities between formal analyses in Chapter 5 and results of principal component analyses. I also consider how Q-factoring relates to the matter of subcultures in research on cultural sentiments.

7.1 COMPONENT ANALYSES

Table 7.1 shows minimum, maximum, and median eigenvalues obtained in component analyses of correlations between respondents' ratings of the 100 stimuli in each of the 15 different sets of stimuli—separate for Evaluation, Potency, and Activity. The table also gives the minimum, maximum, and median ratios of adjacent eigenvalues. Consider the Evaluation dimension first. In data for every stimuli set, the first eigenvalue was far bigger than the second—more than 17 times as big at minimum, and almost 30 times as big at

TABLE 7.1 Eigenvalues Obtained in Principal Components Analyses of Correlations Between Respondents' Ratings of 100 Stimuli in 15 Stimuli Sets

	Eigenvalue				Ratio		
	1	2	3	4	1 to 2	2 to 3	3 to 4
			Evaluation				
Minimum	33.13	1.51	1.41	1.31	17.60	1.02	1.02
Maximum	50.42	2.69	2.06	1.96	29.66	1.38	1.15
Median	43.65	1.89	1.74	1.61	22.14	1.09	1.07
			Potency				
Minimum	18.74	3.93	2.26	2.01	2.70	1.24	1.09
Maximum	29.25	7.75	3.51	2.52	7.21	2.79	1.39
Median	25.33	4.39	2.75	2.27	5.01	1.92	1.19
			Activity				
Minimum	18.30	2.68	2.09	1.93	3.17	1.16	1.05
Maximum	33.22	6.54	3.14	2.74	9.05	2.80	1.26
Median	23.81	4.44	2.37	2.13	5.48	1.89	1.13

Note. The number of respondents rating a stimuli set varied from 64 to 88, with a median of 73.

maximum. The median ratio of first to second eigenvalues was 22.14. The major shifts in size from first to second eigenvalues provide clear evidence that a dominant factor governed the Evaluation judgments made by all respondents across all stimuli sets. Moreover, the second eigenvalue was only a small amount bigger than the third eigenvalue in all 15 analyses—at most 1.38 times as big—and the third was only slightly bigger than the fourth. In all 15 analyses, when eigenvalues after the first were graphed in order of extraction, a smooth, almost flat line connected the points. The small, gradually declining eigenvalues after the first provide clear evidence that just a single common factor governed Evaluation judgments of all respondents, across all stimuli sets.

In the case of the Potency dimension, the first eigenvalue was notably larger than the second in all 15 analyses, although the discrepancy was less than in the case of Evaluation. The first eigenvalue was 2.7 times larger than the second at minimum, and more than seven times as large at maximum, with a median ratio of 5.01. This provides plausible evidence that a dominant factor governs Potency judgments. On the other hand, the ratios of the second to third eigenvalues often were close to 2 (median value: 1.92), raising a question of whether more than one factor was influencing Potency judgments. The third eigenvalues, though, were only slightly larger than the fourth (median ratio: 1.19), and subsequent eigenvalues aligned in a smooth, almost flat line when they were graphed, indicating that at most two factors were consequential in influencing Potency judgments.

Component analyses of Activity data yielded results similar to those obtained on the Potency dimension. The first eigenvalue was notably larger than the second in all 15 analyses: three times as large at minimum, nine times as large at maximum, with a median ratio of 5.48. The ratios of second to third eigenvalues often were close to 2 (median value: 1.89), while ratios of third to fourth eigenvalues were near 1.0 (median value: 1.13). Eigenvalues after the second aligned in a smooth, almost flat line when they were graphed. Thus, at least one common factor influenced Activity judgments, a second factor also may have been operative, but no additional factors were consequential.

Loadings on the second Potency factor correlated with loadings on the second Activity factor, 0.52. Thirty-five percent of respondents had negative loadings on both of the second factors, 32% had positive loadings on both, and 33% had a negative loading on one of the second factors but a positive loading on the other. Assuming that two factors affected respondents' ratings in the case of Potency, I computed the means of Potency ratings among respondents with low loadings on Potency component 2 and also the means for respondents with high loadings on Potency component 2. I then compared the mean Potency scores in the two groups for various concepts. A considerable number of these means were in opposite directions. For example, a gunman had a mean Potency rating of 3.61 in one group and –2.39 in the other group; being violent had a Potency mean of 2.88 in one group and –2.71 in the other; the mean Potency ratings of murdering someone in the two groups were 3.38 versus –4.00; of raping someone 3.61 versus –3.01; of a prison 2.13 versus –1.44; and of a swimming hole –0.27 versus 2.72.

Doing the same thing on the Activity dimension, I found that a terrorist was rated as active by respondents at one extreme on Activity component 2 and inactive by respondents at the other extreme—2.81 versus –1.20; the mean Activity ratings for a guest in the two groups were –1.10 versus 1.27; being considerate had a mean Activity of –1.98 in one group and 1.72 in the other; the Activity of being frightened was 2.37 in one group and –1.49 in the other; whipping someone varied from 2.22 to –3.35; making love to someone from –1.56 to 2.80; a home had an Activity of –1.00 in one group and 3.61 in the other; a church was –1.59 in one group versus 2.54 in the other.

7.1.1 Reduplication

Examination of such differences led to the hypothesis that a subset of respondents skewed their Potency and Activity ratings of a concept in the direction of the concept's Evaluation.[1] To check the hypothesis of Evaluation reduplication in Potency ratings, I computed the correlations of concepts' mean Evaluation ratings, calculated across all respondents, with the concepts' mean

[1]The opposite causal directionality can be ruled out since Evaluations are homogeneous among respondents, aside from levels of inculcation. Thus, concepts' Potencies or Activities cannot be influencing the Evaluations of a subset of respondents.

TABLE 7.2 Correlations Between Mean Evaluation Scores and Mean Potency or Activity Scores, Computed Separately for Respondents Loading Low, Medium, or High on Component 2

	Trichotomized Loadings on Component 2 of:					
	Potency			Activity		
	Lowest	Middle	Highest	Lowest	Middle	Highest
Computations over 100 Concepts in 15 Stimuli Sets						
Minimum correlation	0.03	0.42	0.81	–0.19	0.03	0.43
Maximum correlation	0.45	0.75	0.93	0.13	0.44	0.83
Median correlation	0.34	0.66	0.89	–0.03	0.25	0.69
Computations over 1500 Concepts in Pooled Stimuli Sets						
Correlation	0.28	0.62	0.88	-0.04	0.25	0.66
Proportion of shared variance	0.08	0.38	0.77	0.00	0.06	0.44

Potency ratings, calculated within the lowest, middle, and highest groups of respondents, in terms of the respondents' loadings on Potency component 2.[2]

The first numeric column of Table 7.2 shows the results for the third of respondents having the lowest loadings on Potency component 2. Among these respondents, the correlation was as low as 0.03 within one of the 15 sets of 100 stimuli; the correlation was as high as 0.45 among respondents receiving another of the stimuli sets; and the median correlation in all 15 stimuli sets was 0.34. Carrying out the computation across all 1,500 concepts in the 15 stimuli sets combined, the Evaluation–Potency correlation was 0.28.

These results contrast with those obtained for the third of respondents, with the highest loadings on Potency component 2 (the third numeric column of Table 7.2). Among these respondents, 0.81 was the lowest Evaluation–Potency correlation obtained for any stimuli set, 0.93 was the highest, and 0.89 was the median. The Evaluation–Potency correlation computed across all 1,500 concepts was 0.88. Thus, respondents with a low loading on Potency component 2 had a relatively small correlation between their Evaluation and Potency ratings, while respondents with a high loading on Potency component 2 had a large correlation between their Evaluation and Potency ratings.

Results for the third of respondents having middling loadings on Potency component 2 were in between results for the low and high groups. This can

[2]The first step was to align component 2 loadings for respondents so that high loadings would correspond to a propensity for reduplicating evaluations when such a propensity exists. For some analyses, I pooled the component 2 loadings across stimuli sets. The distributions of pooled loadings were bell-shaped for both Potency and Activity, although the Activity distribution had a long tail of Evaluation reduplicators.

be seen best in the bottom row of Table 7.2, which shows the proportion of shared variance (r^2) in mean Evaluation and Potency ratings. The proportion of shared variance within the middle group—0.38—was about equidistant from the proportions within the lowest and highest groups of respondents, although a bit closer to the lowest group. This indicates that a propensity for reduplicating Evaluation in Potency was a characteristic that varied continuously among respondents rather than a feature of a discrete subgroup of respondents.

I graphed concepts on Evaluation and Potency axes for the third of respondents having the lowest loadings on Potency component 2. Negatively evaluated concepts ranged from high levels of potency for nouns and behaviors related to violence, such as murderer, battlefield, rape, and torture, to low levels of potency for oppressed persons, such as victim and unemployed person and for such personal states as being suicidal or meek.[3] The majority of positively evaluated concepts were rated potent. However, various kinds of children, elders, and disabled persons appear in the positive Evaluation and negative Potency quadrant.

A parallel graph for the third of respondents having the highest loadings on Potency component 2 looked very different. Almost all of the concepts lay on a diagonal band ranging from bad and weak to good and strong, so concepts with a given level of evaluation varied only a little in potency. Outliers were rare: a few concepts of violence, such as mobster and battlefield, were negatively evaluated and viewed as potent; and infant and toddler were positively evaluated while being viewed as impotent.

I followed the same procedures with regard to Activity ratings in order to check the hypothesis of Evaluation reduplication in Activity ratings. I correlated concepts' mean Evaluation ratings with the concepts' mean Activity ratings, among respondents with the lowest, middle, and highest loadings on Activity component 2. The fourth numeric column of Table 7.2 shows the results for the third of respondents having the lowest loadings on Activity component 2. Among these respondents, the correlation was as low as –0.19 in one of the 15 sets of 100 stimuli; the correlation was as high as 0.13 in another of the stimuli sets; and the median correlation in all 15 stimuli sets was –0.03. The Evaluation–Potency correlation was –0.04 across all 1,500 concepts.

The results for the third of respondents having the highest loadings on Activity component 2 are displayed in the sixth numeric column of Table 7.2. Among these respondents, 0.43 was the lowest Evaluation–Activity correlation obtained for any stimuli set, 0.83 was the highest, and 0.69 was the median. The Evaluation–Activity correlation computed across all 1,500 concepts was 0.66. So whereas respondents with a low loading on Activity component 2 had no correlation between their Evaluation and Activity ratings, the Evaluation–

[3]Only a few behaviors, such as serve, obey, mumble to, or beg, are rated low in potency.

Activity correlation was moderate among respondents with a high loading on Activity component 2.

Results for the middle third of respondents on Activity component 2 were closer to the lowest group than to the highest group. The proportion of shared variance (r^2) in mean Evaluation and Activity ratings within the middle group was 0.06, compared to 0.00 in the lowest group and 0.44 in the highest group.

A scatter graph of concepts' Activity scores versus their Evaluation scores was almost square in shape when plotted from scores obtained from the third of respondents having the lowest loadings on Activity component 2. A parallel scatter graph with scores from the third of respondents having the highest loadings on Activity component 2 positioned the concepts mainly within a diagonal swath. However, the bad and active quadrant was sparsely populated with such outliers as gunfight, mob, and villain; and the good and quiet quadrant was sparsely populated with such outliers as retiree, library, and feeling relaxed and peaceful.

Summarizing, the somewhat large second eigenvalues in the Potency and Activity component analyses reflect the fact that Potency or Activity ratings among some respondents were moderately predictable from the Evaluations of concepts, whereas other respondents' Potency or Activity ratings were less influenced by how good or bad a concept was, or not influenced at all.

I investigated whether a propensity for reproducing Evaluation in other kinds of ratings could be predicted from characteristics of respondents. The first finding was that propensity for reduplicating Evaluation in Potency ratings and propensity for reduplicating Evaluation in Activity ratings were somewhat related attributes, although not equivalent to one another. The correlation of Potency component 2 loadings with Activity component 2 loadings was 0.52.

I searched for relationships between component 2 loadings and the individual characteristics considered in Chapter 6. Significant relations showed up for gender, race, school, and geographic origin in one-way analyses of variance. However, some of these effects dropped away when all four variables were considered together in a four-way analysis of variance. Specifically, Asian-Americans were more likely to collapse their Potency ratings down to Evaluation, and so were whites in the Business School. Arts and Sciences students—whether whites, blacks, or Hispanics—were least likely to collapse Potency to Evaluation. These statistically significant effects altogether accounted for 5 percent of the variance in Potency component 2 loadings for individuals.

In the case of Activity component 2, Asian-Americans again were most likely to collapse their ratings down to Evaluation, and Business School students of all races also were likely to do so. White Arts and Sciences students and blacks and Hispanics in the Northeast were least likely to collapse Activity to Evaluation. These significant effects accounted for 6 percent of the variance in Activity component 2 loadings for individuals. No other background variables predicted propensity to reduplicate Evaluations in Potency or Activity ratings.

TABLE 7.3 Concepts with Significantly Different Evaluation Means for Respondents Who Reduplicate Evaluation in Their Potency Ratings Versus Non-reduplicators

Value More	Value Less	Condemn Less	Condemn More
	Reduplicators		
girl-Friday, face someone, supervise, hotel room, compete with, brother-in-law, customer, shopper, employee, consultant, girl, collaborate with, playful, glance at	refreshment stand, desire sexually, Hispanic-American	slug someone, hurry someone, devil	injure, cheat on, homesick
	Non-reduplicators		
Disvalue			
middle-aged, military boot camp, subway, assembly line, contradict			

Note. Evaluation means were computed among the third of respondents with the lowest loadings on Potency component 2 and the third with highest loadings. Means in the two groups were significantly different at the 0.01 level (two-tailed test, Student's *t*) for listed concepts. Concepts with larger differences in means are listed before concepts with lesser differences.

While comparing the ratings of reduplicators with ratings from non-reduplicating respondents, I discovered that reduplicators differed in Evaluations for some concepts. Table 7.3 shows concepts that were evaluated differently by respondents who reduplicate Evaluation in their Potency ratings, as opposed to respondents whose Potency ratings were largely independent from Evaluations. Table 7.4 shows concepts that were evaluated differently by respondents who reduplicate Evaluation in their Activity ratings, as opposed to respondents whose Activity ratings were largely independent of Evaluations.

Some of the differences cataloged in Tables 7.3 and 7.4 undoubtedly are sampling errors. Each table is the product of testing mean differences for 1,500 concepts, and one would expect 15 differences to be significant at the 0.01 level purely by chance. However, the 28 entries in Table 7.3 and the 25 entries in Table 7.4 are beyond the numbers expected by chance. Entries at the top of each list probably are more reliable pointers to value differences between groups since the entries are listed with larger differences first. I leave it to the reader to suss out the general nature of reduplication subcultures from the content of Tables 7.3 and 7.4.

Reduplicators may have an ideological perspective in which powerful, or active, things generally are viewed as good, and powerless, or passive, things generally are viewed as bad. For such persons the Potency and Activity scales

TABLE 7.4 Concepts with Significantly Different Evaluation Means for Respondents Who Reduplicate Evaluation in Their Activity Ratings Versus Non-reduplicators

Value More	Value Less	Condemn Less	Condemn More
	Reduplicators		
traveler, kind, relative, grocery store, prod someone, bridesmaid, athlete, factory, uncle, stepbrother, Navy reservist, employee, sports fan, shopper, vacationer, shake hands with	—	contemptuous	gun moll, gangster, widower, halt someone
	Non-reduplicators		
Disvalue self-conscious, vigilante, interrogation, attorney			

Note. Evaluation means were computed among the third of respondents with the lowest loadings on Activity component 2 and the third with highest loadings. Means in the two groups are significantly different at the 0.01 level (two-tailed test, Student's t) for listed concepts. Concepts with larger differences in means are listed before concepts with lesser differences.

used in this study could have turned into de facto Evaluation scales. Assessing the true Potency and Activity sentiments of such people would require use of projective measurements (Raynolds, Sakamoto, and Raynolds 1988; Raynolds, Sakamoto, and Saxe 1981) or instruments whose dimensions are defined graphically rather than verbally, as discussed in Section 2.2.3.

7.1.2 Inculcation and Commonality

Romney, Batchelder, and Weller (1987) stated that informants' levels of cultural inculcation could be obtained by Q-factoring informants' answers to a set of questions. "The factor loadings ... represent the estimated competence of each informant for the cultural domain under consideration" (1987, p. 172). I checked this claim by correlating enculturation indices based on the formulas obtained in Chapter 5 with loadings on component 1 in the Q-factor analyses discussed above. The correlations of inculcation indices with component loadings over 1,113 respondents were 0.82 for Evaluation, 0.81 for Potency, and 0.79 for Activity. Although these correlations are high, they do not indicate that the component loadings are the same thing as inculcation indices. However, the loadings on component 1 correlate much higher with commonality indices: 0.997 on Evaluation, 0.996 on Potency, and 0.995 on Activity.

These figures are sufficiently high to indicate that the component 1 loadings obtained by factoring inter-respondent correlations are equivalent to commonality indices.

Inculcation indices can also be estimated from component 1 loadings in Q-factor analyses, by factoring variances and covariances among respondents instead of correlations. The correlations of inculcation indices with component 1 loadings obtained from component analyses of variances and covariances were 0.97 for Evaluation, 0.98 for Potency, and 0.97 for Activity. These figures are high enough to consider the loadings from principal component analyses of variances and covariances as essentially equivalent to inculcation indices.[4]

Recapitulating, inculcation indices can be estimated by computing the variances and covariances of respondents' answers, deriving the first principal component of the variance–covariance matrix, and interpreting the component loadings of respondents as inculcation indices. Commonality indices can be estimated by computing correlations among respondents' answers, deriving the first principal component of the correlation matrix, and interpreting the component loadings of respondents as commonality indices.

7.2 SUBCULTURES

Subcultures skew sentiments regarding a few concepts while leaving cultural sentiments unaffected for the vast majority of concepts. "Individuals who disagree too much to adopt a society's normative sentiment about something may gravitate to a special group that provides them with better affective resonance. As individuals segregate themselves in this way, diverging pockets of consensus—or subcultures—emerge. Societal diversity in sentiments about an issue often corresponds not to anarchic individuality but to the existence of subcultures. . . . Subcultures orbit around types of people, actions, and material objects that are of special significance within the sub-population. Individuals in the sub-population typically have more positive sentiments about the focal matters than do individuals in the culture at large" (Heise 2007, p. 21).

The existence of subcultures raises the question of how they would be reflected in Q-factoring of sentiment measurements. Suppose that members of the subculture have some sentiments opposite in valence to the sentiments of most people in the population, as occurs in deviance subcultures (Heise 2007, pp. 24–25). Then, if the number of contrasting sentiments is a substantial proportion of the total number of sentiments being measured, an appreciable second principal component will appear when inter-respondent correlations are Q-factored, and the large second component signals the possible existence of a subculture. However, if the number of contrasting sentiments is small

[4]On the other hand, the component 1 loadings from analyses of variances and covariances are not adequate replacements for commonality indices, where the correlations between commonality indices and component 1 loadings are 0.82 on Evaluation, 0.79 on Potency, and 0.78 on Activity.

relative to the total number of sentiments being measured, the subculture would generate only a small principal component, and the subculture might go undetected.

Suppose, instead, that members of the subculture either attenuate or exaggerate cultural sentiments that are of particular relevance to them. For example, police officers exaggerate their own goodness and attenuate criminals' badness relative to widely held cultural sentiments (Heise 1979, p. 100). When Q-factoring correlations among respondents, such a subculture will be reflected in a principal component beyond the first, but the component will be highly correlated with the general culture component and the eigenvalue corresponding to the orthogonal component will be small. The presence of the subculture probably would not be detected in such a case, especially if the number of items of subcultural relevance are few relative to the total number of sentiments being measured.

These ideas can be illustrated with the data at hand. The first case of finding oppositional subcultures through Q-factoring already has been demonstrated in the discovery of subsets of respondents who reduplicate Evaluation in their Potency or Activity ratings, leading to Potency or Activity ratings that often are in opposition to the ratings of non-reduplicators. When Q-factoring Potency and Activity ratings, the reduplication phenomenon created noteworthy second principal components. Reduplication may be interpretable in terms of subcultures, even to the extent that reduplicators and non-reduplicators differ in their evaluations of some concepts as well as in their Potency and Activity ratings of concepts.

The second case involving the feebleness of Q-factoring for detecting subcultures produced by attenuation or exaggeration of general sentiments can be illustrated with gender subcultures.

7.2.1 Gender

Repeated studies have revealed gender differences in sentiments (Heise 2007, p. 22). The differences arise for relatively few concepts, and apart from this limited arena of differentiation, females and males have about the same sentiments. Thus, gender offers an example of subcultures.

I compared the mean sentiment ratings of females and males for the 1,500 concepts considered in this study, using a stringent criterion of significance to sift out the most notable concepts that distinguish female and male sentiments. Table 7.5 lists the concepts for which gender differences in sentiments were significant at the 0.001 level in two-tailed tests using Student's t.

The table shows that more than males do, females condemn violence (gunfight, hurting, clubbing, slaughterhouse, slugging) and unrestrained sexuality (whorehouse, following, peeping at), while approving more of femaleness (female, feminist, feminine), female concerns (boyfriend, beauty salon), and concepts related to affiliation (roommate, relative, restaurant). Among character traits, females are more disapproving of self-consciousness and

TABLE 7.5 Concepts with Male–Female Differences in Sentiments

Females More Negative	Females More Positive
	Evaluation
whorehouse, gunfight, hurting, self-conscious, clubbing, slaughterhouse, slugging, following, peeping at	female, boyfriend, beauty salon, roommate, relative, feminist, feminine, restaurant, gentle
	Potency
modest	rapist, female, feminine, sermon, grading
	Activity
underdog, self-conscious, glum	saloon, head nurse, foster mother

Note. Results show concepts where males and females are significantly different at the 0.001 level in two-tailed tests based on Student's t statistic, conducted across 1,500 concepts. Concepts are listed in each cell in the order of the male–female differences, biggest differences first.

approving of gentleness. Females see a sexual predator (rapist) as more potent than males do, as well as femaleness (female, feminine), and some standard institutional activities (sermon, grading). They see the character trait of modesty as more impotent than males do. Females see a saloon as more active than males do, and also some female authority figures (head nurse and foster mother), while viewing an underdog as more passive than males do. The character traits of self-consciousness and glumness also are felt to be more passive by females than by males.

While one or two differences significant at the 0.001 level might occur purely by chance when examining 1,500 instances, the existence of 30 significant differences on the three affective dimensions demonstrates that gender differences in sentiments are a reality in middle-class American culture.[5] On the other hand, the two genders do not differ in sentiments for most concepts. Eighty-one percent of the 1,500 concepts have no significant gender difference at the 0.05 level on any of the EPA dimensions.

Within-gender consensus about concepts like those shown in Table 7.5 undoubtedly augmented correlations among females and correlations among males, compared to intergender correlations. Yet the increases in levels of correlations within genders were too meager to be reflected in Q-factoring. This was especially obvious in the analyses of Evaluation ratings, where the first principal components always were very large and all other principal components always were insubstantial. Some of the insubstantial components in Evaluation analyses probably corresponded to female and male patterns, but those principal components were so small that they seemed to be no more

[5] Such gender differences also occur in some other societies (Heise 2007, p. 22).

than statistical litter. Similarly, the Q-analyses of Potency and Activity also probably resulted in principal components related to gender, but the components were too trivial to warrant investigation.

7.3 DISCUSSION

Q-factoring fractionates the similarities among respondents' stimuli responses into a set of principal components. Finding a single dominant principal component supports the basic assumptions of culture-as-consensus methodology: that respondents are drawn from a common culture generating similarity in responses because all respondents conform to relevant cultural norms to a greater or lesser degree.

The analyses in this chapter demonstrated that the assumptions of culture-as-consensus methodology are satisfied quite well in the case of Evaluation ratings: Q-factoring produced a massive first principal component and negligible additional components in all analyses. Analyses provided moderate support for the assumptions with regard to Potency and Activity ratings: a dominant principal component appeared in all analyses, yet appearances of a small but noteworthy second component suggested that two different factors often were in play. The second component, it turned out, resulted from some respondents reduplicating evaluations of stimuli in their Potency or Activity ratings.

Reduplicating Evaluation in other kinds of ratings could be an individual idiosyncrasy, but this propensity was shared among a subset of respondents, leading to a suspicion that subcultures were involved. That conjecture was buttressed by finding some significantly different evaluations among reduplicators compared to non-reduplicators, in addition to the differences in Potency or Activity ratings.

Reduplication phenomena appear to be ubiquitous since they were evident in every subsample in this study, and reduplication processes also appeared in data collected about two decades earlier (Thomas and Heise 1995, Figure 2). Because reduplication subcultures have not been identified previously, sentiment measures up to now have been computed by averaging across all respondents—reduplicators and non-reduplicators alike—thereby attenuating estimates of some Potency and Activity sentiments and manufacturing inflated correlations among EPA measurements. EPA assessments and their applications perhaps can be improved by creating separate databases of sentiments for reduplicators and non-reduplicators, as is done currently with regard to gender. Q-factor loadings on the second principal component could serve as indices to identify reduplicators, or special items might be included in surveys to identify them (e.g., is a gunman strong or weak?; is a terrorist lively or quiet?; is a church lively or quiet?).

Q-factoring offers an accessible and relatively straightforward method for obtaining inculcation and commonality indices. In Chapter 6, I implemented the formulas derived in Chapter 5 with the aid of mathematical software that is uncommon in social researchers' toolkits. However, results in this chapter show that indices equivalent to those computed from the formulas can be obtained with standard statistical packages. Specifically, inculcation indices for respondents can be obtained by Q-factoring the matrix of covariances among respondents' answers, and commonality indices can be obtained by Q-factoring the matrix of correlations among respondents' answers. Respondents' loadings on the first principal component of a covariance factoring constitute their inculcation indices and loadings on the first principal component of a correlation factoring constitute their commonality indices. While Q-factoring is not a direct option in statistical packages when data are in the standard form of a record for each respondent, commands can be marshaled to accomplish the task. For example, Q-factoring in SPSS involves first applying the FLIP VARIABLES function to turn responses into cases and respondents into variables, and then invoking the FACTOR function.

As stated in Chapter 1, respondents in surveys of culture should be people who reproduce the culture of interest and who are to be found in the behavior settings where the culture is being reproduced. The empirical study analyzed in this book targeted students on college campuses as reproducers of middle-class American culture. Such respondents were presumed to be fairly homogeneous with regard to sentiments, and the Q-factorings of their responses indicate that this was the case, notwithstanding demographic variations in gender, race–ethnicity, and geographic origin.

The unidimensionality of respondents' evaluation ratings are particularly provocative. On the whole, respondents had the same patterns of evaluations across concepts, with individual respondents varying primarily in their overall tendency to attenuate or exaggerate general evaluation norms. An implication is that many attitudes and opinions of college students are set by culture, and individual differences are associated largely with the extent to which individuals inculcate cultural norms. This contrasts with viewing individual variations as independent and random—a model that does apply sometimes (e.g., in males' opinions about birth control methods, Romney, Batchelder, and Weller 1987, p. 175) but that evidently is not the appropriate model for opinions about many things.

7.4 CHAPTER HIGHLIGHTS

- A single set of norms organized respondents' Evaluations of 100 concepts. The ratings of most respondents correlated substantially with the norms, and respondents with low commonality shared no alternative Evaluations of concepts that were substantial enough to show up as a secondary component in factor analyses.
- Potency and Activity ratings were organized by two different kinds of norms. About a third of the respondents provided Potency ratings that

were largely independent of Evaluation or Activity ratings, and provided Activity ratings that were largely independent of Evaluation or Potency ratings. However, the remaining respondents reduplicated Evaluation norms in their Potency or Activity ratings.

- About half of the respondents with a propensity to reduplicate Evaluation did so in both Potency and Activity ratings, while the other half reduplicated Evaluation in Potency ratings or in Activity ratings but not both. Asian-Americans and Business School students were more likely than others to reduplicate Evaluation in their ratings of Potency or Activity.
- Reduplicators and non-reduplicators differed in their evaluations of some concepts as well as in their Potency and Activity ratings of concepts. Thus, reduplication of evaluation norms may be part of a subculture.
- Females condemned violence, unrestrained sexuality, and self-consciousness more than males did, while approving more of femaleness, affiliation, and gentleness. More than males, females felt that sexual predation is potent, and modesty is impotent; and they felt that some female authority figures are active, while an underdog, self-consciousness, and glumness are passive. Such differences in sentiments notwithstanding, component analyses failed to reveal gender subcultures because the differences that occurred largely amounted to normative attenuation by one gender or the other, as if one gender was less inculcated regarding particular cultural sentiments.
- Inculcation of respondents into the normative culture can be estimated by deriving the principal components of the variance–covariance matrix of their ratings computed across concepts: loadings on the first principal component index inculcation. The commonality of respondents' answers can be assessed by deriving the principal components of the correlation matrix of their ratings computed across concepts: loadings on the first principal component index commonality.

8 Measurement Reliability

Reliability coefficients estimate the proportions of meaningful variations in measurements, as opposed to transitory, unaccountable variations such as those from measurement errors. Alwin (2007) estimated the reliabilities of various kinds of survey questions as well as the impact of various factors impinging on measurement reliability, such as the education and age of respondents, whether respondents were reporting about themselves or others, whether they were reporting facts or subjective variables, and where a question appeared within a questionnaire. Being interested in the reliabilities of single survey questions rather than of tests based on multiple items, Alwin assessed reliability from repeated measures administered at three different times. The three-wave data allowed Alwin to separate unreliability and instability with models introduced by Heise (1969b), Wiley and Wiley (1970), and Werts, Jöreskog, and Linn (1971). Among other things, Alwin (2007) found that open-ended questions are more reliable than closed-ended questions, that questions with briefer texts are more reliable than wordy questions, and that the number of response categories in closed-ended questions relates to reliability in complicated ways.

Of particular interest in this book, Alwin (2007, Table 9.3) analyzed the reliabilities of 47 "feeling thermometers." Feeling thermometers are bipolar scales, presented visually in a questionnaire as a picture of a mercury thermometer, and used to measure subjective feelings of warmth versus coldness toward various stimuli. Feeling thermometers are a standard method for measuring sentiments in surveys, rather than rating scales like those presented in Chapter 2, which require more technological overhead.

Alwin found that the average reliability of feeling thermometers was 0.65. Alwin's reliability estimate, derived in a methodologically sophisticated way, sets an expectation about the reliabilities of the bipolar rating scales used in this book. The Alwin estimate suggests that about 65 percent of the variance in ratings on sentiment-rating scales of a single stimulus by individual respondents is reliable and repeatable, and 35 percent of the variance is unaccountable disturbance that changes from one time to another.

Computerized bipolar rating scales currently used to measure sentiments are more precise than feeling thermometers [which in practice get used as nine-

Surveying Cultures: Discovering Shared Conceptions and Sentiments, By David R. Heise

position scales according to Alwin (2007, p. 189)]. Moreover, Alwin's estimates of reliability overlook the three-dimensional nature of sentiments, providing a single assessment of reliability when three different reliabilities might be more appropriate. Thus, in this chapter I supplement Alwin's studies by assessing the reliabilities of the contemporary sentiment measuring instrument presented in Chapter 2, for Evaluation, Potency, and Activity measurements.

I first compute traditional reliabilities in the form of test–retest correlations for one stimulus at a time; individual variations in sentiments are what constitute "meaningful variations" in this case. I then recompute reliabilities across stimuli, and I treat variance associated with stimuli as "meaningful variations." As expected from formal analyses in Chapter 5, traditional reliabilities are low for norm measurements, but reliabilities are higher when viewing variation across stimuli as the meaningful aspect of sentiment ratings. The chapter ends by showing that averaged ratings from multiple respondents provide norm measurements of high reliability.

8.1 RELIABILITIES WITHIN STIMULI

The test–retest data that I use in this chapter are from the Business School subsample within the project described in Chapter 3. Time 1 measurements corresponded to the standard survey in which each respondent rated a set of 100 stimuli, with each set selected randomly from 15 different sets. In a second survey about six weeks later every respondent rated eight identities and eight behaviors that had been in various sets of stimuli in the first survey. The ratings of the 16 stimuli constitute time 2 measures. Identifying information collected from respondents in both surveys allowed the time 1 and time 2 ratings to be linked.

Table 8.1 summarizes the test–retest data. Statistics on identities are given in part A of the table; statistics on behaviors are given in part B. The column in Table 8.1 titled *Stimulus* shows which identity or behavior was rated and whether the statistics in the row are for Evaluation, Potency, or Activity ratings.

The columns under *Time 1* give the means and variances of ratings and the numbers of people who rated the stimulus during the first survey. For example, 56 people rated "a chain smoker" at time 1, and the mean of their Evaluation ratings was –2.22, with a variance of 4.11. The columns under *Time 2* give the means, variances, and numbers of repeat raters for the second survey. The repeat raters were those who had been presented with the given stimulus at time 1. Small discrepancies occur in sample sizes since some respondents skipped a stimulus at time 1 but not time 2, while others skipped a stimulus at time 2 but not time 1.

The column titled *N* under *T1, T2* gives the number of people who rated each stimulus at both time 1 and time 2. The column titled *r* shows product-moment correlations between individuals' time 1 and time 2 ratings.

TABLE 8.1 Statistics for Evaluation (E), Potency (P), and Activity (A) Ratings of Identities and Behaviors

	Time 1			Time 2			T1, T2		All		
Stimulus	N	Mn	Var.	N	Mn	Var.	*N*	*r*	N	Mn	Var.
				A. Statistics for Identities							
chain smoker E	56	−2.22	4.11	54	−2.54	2.87	54	0.35	712	−2.19	3.47
chain smoker P	56	−1.60	4.81	54	−1.79	3.50	54	0.57	712	−1.54	3.72
chain smoker A	56	−0.66	6.24	54	−0.97	4.69	54	0.25	712	−0.97	4.61
executioner E	55	−2.28	2.49	57	−2.18	2.84	54	0.38	706	−1.75	3.71
executioner P	55	1.86	5.76	57	2.04	4.78	54	0.54	706	1.68	4.16
executioner A	55	−0.90	5.27	57	−0.22	6.43	54	0.67	706	−0.34	5.33
intern E	41	1.74	3.35	42	1.50	2.70	41	0.41	719	1.57	2.40
intern P	41	−0.56	3.56	42	−0.03	4.31	41	0.42	719	−0.46	4.10
intern A	41	0.61	4.16	42	0.91	3.34	41	0.49	719	0.71	3.01
racketeer E	35	−0.52	3.54	35	−0.16	2.65	29	0.62	619	−0.50	2.18
racketeer P	35	0.91	2.22	35	1.04	2.39	29	0.61	619	0.52	2.14
racketeer A	35	1.30	1.98	35	0.68	3.40	29	0.32	619	0.77	2.17
retiree E	40	1.29	3.01	40	1.72	1.76	40	0.29	707	1.43	2.30
retiree P	40	0.14	3.54	40	0.07	2.81	40	0.12	707	−0.20	2.79
retiree A	40	−1.01	3.72	40	−1.18	2.64	40	0.14	707	−1.17	2.43
salesman E	50	−0.04	3.19	49	0.20	2.76	49	0.21	713	0.29	2.50
salesman P	50	0.60	2.42	49	0.93	2.79	49	0.44	713	0.66	2.45
salesman A	50	1.79	2.32	49	1.78	2.29	49	0.49	713	1.59	2.54
scientist E	49	1.77	2.40	49	1.44	2.33	48	−0.01	717	1.49	2.36
scientist P	49	1.72	2.58	49	1.44	3.37	48	0.27	717	1.48	2.59
scientist A	49	−0.11	3.84	49	0.00	3.20	48	0.53	717	−0.23	3.29
teammate E	40	2.00	2.15	40	1.56	3.72	40	0.27	718	1.90	2.32
teammate P	40	1.56	2.20	40	1.26	2.41	40	0.12	718	1.46	2.19
teammate A	40	1.42	2.20	40	1.26	3.26	40	0.23	718	1.30	2.19

TABLE 8.1 (*Continued*)

	Time 1			Time 2			T1, T2		All		
Stimulus	N	Mn	Var.	N	Mn	Var.	*N*	*r*	N	Mn	Var.
				B. Statistics for Behaviors							
bickering with E	51	−1.42	2.02	53	−1.73	2.07	51	0.33	709	−1.41	2.19
bickering with P	51	−0.24	2.92	53	−0.40	3.17	51	0.40	709	−0.14	2.99
bickering with A	51	1.00	2.87	53	1.56	3.00	51	0.10	709	1.23	2.67
billing E	53	−0.68	2.46	55	−0.41	1.86	53	0.49	703	−0.09	2.19
billing P	53	0.99	2.05	55	0.75	2.47	53	0.21	703	0.86	2.21
billing A	53	−0.56	1.92	55	−0.22	1.78	53	0.44	703	0.03	2.19
bossing around E	54	−1.60	3.00	53	−0.89	2.96	53	0.42	714	−1.14	2.80
bossing around P	54	0.86	3.95	53	0.92	3.52	53	0.36	714	0.87	3.82
bossing around A	54	0.85	2.97	53	1.21	2.71	53	0.06	714	1.17	2.53
murmuring to E	39	−0.61	2.89	41	−0.59	1.84	38	0.28	700	−0.48	1.82
murmuring to P	39	−0.98	2.13	41	−1.01	1.64	38	0.27	700	−0.76	1.83
murmuring to A	39	−1.51	2.31	41	−1.58	1.44	38	0.16	700	−1.09	2.04
requesting from E	40	0.70	2.21	40	0.55	1.92	40	0.20	713	0.57	1.78
requesting from P	40	−0.01	3.49	40	0.03	2.94	40	0.53	713	0.07	2.39
requesting from A	40	−0.07	2.30	40	−0.04	2.24	40	0.16	713	0.19	1.96
selling to E	44	1.41	1.94	43	1.44	1.41	43	0.38	714	1.19	1.74
selling to P	44	1.51	2.39	43	1.42	1.46	43	0.20	714	1.11	2.05
selling to A	44	1.45	2.31	43	1.33	1.77	43	0.47	714	1.02	2.13
waiting on E	50	0.56	3.94	51	0.25	4.39	50	0.40	713	0.13	3.68
waiting on P	50	−0.14	3.24	51	−0.18	2.74	50	0.48	713	−0.47	2.87
waiting on A	50	−0.60	3.70	51	−0.44	3.84	50	−0.19	713	−0.55	3.21
whispering to E	45	0.79	1.72	44	0.22	2.40	44	0.26	712	0.34	2.25
whispering to P	45	0.20	2.10	44	−0.41	3.07	44	0.13	712	−0.28	2.50
whispering to A	45	−1.56	2.35	44	−1.25	3.13	44	0.35	712	−1.37	2.62

Note: Abbreviations: Mean: Mn; Variance: Var; All respondents at T2: All.

The columns titled *All respondents at T2* show the means and variances of ratings and the numbers of people who rated the stimulus during the second survey when the 16 stimuli in Table 8.1 were presented to every respondent. The total number of respondents retained at time 2 was 722, and the difference of each *N* from 722 equals the number who skipped the stimulus.

Table 8.1 warrants some comments.

- The means of *All respondents at T2* reveal that the attempt to represent all eight types of sentiments in the test–retest study (see Table 3.1) was achieved imperfectly: a bad, weak, lively identity did not materialize; nor did good behaviors that are either powerful and quiet, or powerless and lively. Nonetheless, the means of the stimuli do have moderate range on each of the EPA dimensions.
- On the whole, Evaluation rating variances are larger at time 1 than at time 2—a sign test of the differences is significant at the 0.05 level when considering the 16 differences in variances of Evaluation ratings for identities and behaviors combined. However, no significant result was found with sign tests structured in other ways (e.g., differences in Evaluation variances within identities or within behaviors, or differences in Potency variances or in Activity variances).
- The *N*s for *All respondents at T2* show that in general 3 percent or less of the respondents skipped any given stimulus. The stimulus "racketeer" was the exception, suggesting that the concept of racketeer was unfamiliar for many of the respondents.
- A substantial number of correlations between time 1 and time 2 ratings were significantly greater than zero—20 of the 24 identity correlations, and 16 of the 24 behavior correlations. (In Table 8.1 correlations of 0.21 or more are significant with $p \leq 0.05$ in a one-tailed test.) The median over-time correlation was 0.36 for identities and 0.30 for behaviors. Combining identities and behaviors, the median correlations were 0.34 on Evaluation, 0.38 on Potency, and 0.28 on Activity. The largest over-time correlation was 0.67 for activity ratings of "executioner."

The test–retest correlations in Table 8.1 constitute reliability coefficients, and their magnitudes are far below the reliability estimates discussed in the introduction to this chapter—the median values of the test–retest correlations are around half the average reliability of feeling thermometers. Next, I amalgamate reliability information for different concepts into an overall summary measure.

Figure 8.1 shows a causal model for respondents' ratings of one stimulus on one dimension at two different times. The model assumes that substantive changes in cultural sentiments are negligible over short periods. MacKinnon and Luke (2002, p. 311) examined changes in sentiments over 14 years for 102 identities in Canadian society and found that "1981 values explain between

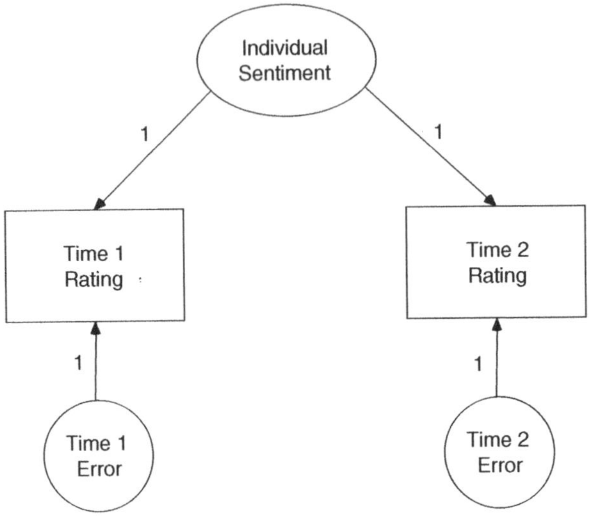

FIGURE 8.1 Causal diagram for analyzing test–retest data.

78% to 80% of the variance of 1995 EPA values for males and between 75% to 84% for females. Thus … cultural sentiments for social identities are relatively stable over time." While MacKinnon and Luke did find significant changes for some specific concepts, those were concepts that had been in play within Canadian media or politics during the intervening years. Interpreting Figure 8.1, a rating at time 1 (Time 1 Rating) is produced from the respondent's sentiment regarding the concept (Individual Sentiment) and from haphazard factors operative at the moment (Time 1 Error). Similarly, the rating at time 2 is a function of the respondent's persistent sentiment regarding the concept plus haphazard factors at time 2. The error variable is assumed to have an average value of zero, and errors are assumed to be uncorrelated over time.

Parameters of the model shown in Figure 8.1 were estimated in 15 different groups simultaneously, each group constituting a different subsample of the 722 respondents in the test–retest analyses. In 14 of the groups, the respondents provided time 1 and time 2 ratings of one stimulus listed in Tables 3.1 and 6.1. Respondents in the fifteenth group rated two stimuli (retiree and requesting) at both times, so for that group the causal diagram in Figure 8.1 is doubled, with allowance for a possible correlation[1] between individual sentiments regarding the two concepts. A structural equation modeling program (Arbuckle and Wothke 1999) was used to obtain maximum-likelihood esti-

[1]Depending on the model being estimated, the correlation is 0.10 or less for Evaluation data, 0.46 or less in the case of Potency, and 0.36 or less with Activity.

mates of variances, covariances, and means on each EPA dimension, with maximum-likelihood adjustments for skipped ratings.[2]

For each dimension, the estimations were done three times with increasingly constrained models. All models assumed that measurement-error variance remained constant over periods of a few months; thus, time 1 measurement errors were constrained to have the same variance as time 2 errors [however, see the thoughtful discussion of this issue by Alwin (2007, p. 107)]. Additionally, all models assumed that no significant changes in sentiment norms occurred over an interval of a few months; thus, the means of ratings for each concept were constrained to be equal across the two times.

- *Stable measurement-error variances.* With this model, measurement-error variances were allowed to be different for different concepts. Variances associated with individual differences in sentiments were free to vary from one concept to another.
- *Stable and equal measurement-error variances.* In this model a single measurement-error variance applied at both times for all concepts, but variances of individual differences in sentiments were free to vary from one concept to another.
- *Stable and equal variances.* In this model a single measurement-error variance applied throughout and individual-difference variances were constrained to be the same for all concepts.

Table 8.2 shows some goodness-of-fit statistics for the three different estimations. The first two numerical columns show the values of chi-square and the degrees of freedom for the model in that row; chi-square is derived from discrepancies between observations and expectations based on the model. Next is the probability that such a large chi-square would result with the given degrees of freedom. The fourth numerical column shows the result of dividing chi-square by the degrees of freedom—here called the CDF ratio for brevity. Kline (1998, p. 128) forwards a suggestion that the CDF ratio should be less than 3.0 for an acceptable model; however, texts can be found suggesting that the critical ratio might go as high as 5.0 or as low as 2.0 (Arbuckle and Wothke 1999, pp. 399–400). The fifth numeric column in Table 8.2 presents the Browne–Cudeck criterion (BCC), which assesses discrepancies between

[2]I recomputed all statistics and parameter estimates with a subsample purged of 85 respondents who lacked ratings at one or the other time, and nine additional respondents who took 13 minutes or less for the time 1 survey. Test–retest correlations in Table 8.1 were somewhat larger, when affected at all, with a maximum increase of 0.09. Model fit statistics paralleled those reported in Table 8.2 and resulted in no changes in decisions about the best model. Estimates of error variance in Table 8.3 were the same for Activity, 0.01 less for Potency, and 0.02 less for Evaluation. Estimates of Evaluation and Potency individual variance were notably larger for *racketeer*, but otherwise fairly close to the values shown in Table 8.3.

TABLE 8.2 Summary Statistics from Structural Equation Analyses with Different Constraints on Variances and Means in Test–Retest Ratings of 16 Concepts

	Chi^2	DF[a]	P	Chi^2/DF	BCC[b]
	Evaluation				
Stable error variances	50.053	35	0.05	1.430	156.120
Stable and equal error variances	74.438	50	0.01	1.489	148.036
Stable and equal variances	101.742	65	0.00	1.565	142.870
	Potency				
Stable error variances	26.987	35	0.83	0.770	133.054
Stable and equal error variances	49.092	50	0.51	0.982	122.689
Stable and equal variances	98.079	65	0.01	1.509	139.207
	Activity				
Stable error variances	38.256	35	0.32	1.093	144.322
Stable and equal error variances	103.982	51[c]	0.00	2.080	177.580
Stable and equal variances	169.162	66	0.00	2.563	208.125

[a]Degrees of freedom.
[b]Browne–Cudeck criterion.
[c]A negative variance estimate arose with this model; that variance was set equal to zero, increasing degrees of freedom by one.

model predictions and observed data, and simultaneously takes parsimony into account. No value of the Browne–Cudeck criterion is associated with rejection of a model. Rather, the index is used to compare different models, giving simultaneous consideration to both goodness of fit and simplicity.

I now turn to an examination of the statistics in Table 8.2 in order to assess the value of each different estimation as an interpretation of the data. In the case of Evaluation ratings, the chi-square probabilities indicate that no model was satisfactory. The CDF ratio suggests that any of the three models amounted to an adequate interpretation of the data, and the BCC indexes indicate that the model with "stable and equal variances" was the best choice, considering its simplicity.

In the case of Potency ratings, the chi-square probabilities indicate that both the "stable measurement-error variances" and the "stable and equal measurement-error variances" models sufficed, the CDF ratio suggests that any of the three models amounted to an adequate interpretation of the data, and the BCC indexes indicate that the model with "stable and equal variances" is the best choice, in that it had low discrepancy for its relative simplicity. With Activity ratings all criteria suggest that the "stable measurement-error variances" model is best. However, all three models might be acceptable according to the CDF ratio.

Giving emphasis to parsimony, the "stable and equal measurement-error variances" model is the best overall interpretation of the data for all three

EPA dimensions. However, the most constrained model, "stable and equal variances," also is a plausible interpretation of the data for Evaluation and Potency and may be an adequate interpretation for some purposes in the case of Activity.

Table 8.3 shows estimates of variances[3] obtained with the two most constrained models considered in Table 8.2. Estimates of the general measurement-error variance are the same in both models, to two decimal places. Thus, the models differ only in allowing individual variances to differ from one concept to another, as opposed to imposing an average individual variance for all concepts.

Table 8.3 warrants a number of comments.

TABLE 8.3 Estimates of Individual Variances and Error Variances

	Evaluation		Potency		Activity	
	Error	Individual	Error	Individual	Error	Individual
	Model with Stable and Equal Error Variances					
chain smoker	1.81	1.40	1.91	2.23	2.22	2.23
executioner	1.81	0.91	1.91	3.06	2.22	3.67
intern	1.81	1.17	1.91	1.77	2.22	1.59
racketeer	1.81	1.42	1.91	0.77	2.22	0.64
retiree	1.81	0.59	1.91	0.78	2.22	0.65
salesman	1.81	0.87	1.91	0.87	2.22	0.58
scientist	1.81	0.25	1.91	0.89	2.22	1.55
teammate	1.81	0.89	1.91	0.30	2.22	0.53
bickering	1.81	0.42	1.91	1.13	2.22	0.48
billing	1.81	0.66	1.91	0.39	2.22	0.19
bossing	1.81	1.16	1.91	1.52	2.22	0.37
murmuring	1.81	0.54	1.91	0.21	2.22	0.00
requesting	1.81	0.31	1.91	1.44	2.22	0.17
selling	1.81	0.22	1.91	0.17	2.22	0.36
waiting on	1.81	1.94	1.91	1.21	2.22	0.40
whispering	1.81	0.36	1.91	0.47	2.22	0.69
	Model with Stable and Equal Variances					
	1.81	0.83	1.91	1.12	2.22	0.93

Note. Individual Activity variance for "murmuring" was constrained to zero in the Equal-Error-Variances-Stable-Means model, because the solution without this constraint produced a negative value for the variance.

[3]Table 8.3 does not include the maximum-likelihood estimates of means because substantive sentiments are not a focus in this study. The interested reader can average the time 1 and time 2 means in Table 8.1 to approximate the maximum-likelihood estimates of EPA profiles for each concept.

- Taking square roots of the measurement-error variances to convert to standard deviations, we get 1.35 on Evaluation, 1.38 on Potency, and 1.49 on Activity. From these values and an assumption of normality, we can infer that a third or more of the EPA ratings that a respondent gives are more than 1.4 scale units away from the respondent's true sentiment—this on a scale with a range of 9.6.
- Some of the indexes of fit in Table 8.2 favor the most constrained model for Evaluation ratings, which suggests that an individual variance of 0.83 applies for all concepts. On the other hand, Table 8.3 does reveal some change in individual variances across concepts when the variances are unconstrained. For example, almost no individual differences arise among these Business School respondents in evaluations of "selling something to someone," whereas individual differences are a bigger source of variance than measurement errors in evaluating "waiting on someone."
- An individual variance of 1.12 can be applied on the Potency dimension for all concepts according to the CDF index of fit. However, freeing individual variances yields the preferred model in the case of Potency, according to goodness-of-fit statistics. The reason is evident in Table 8.3: Almost no individual differences arise in assessing the potency of "murmuring to someone" (a variance of 0.21), but an individual variance of 3.06 occurs in assessing the powerfulness of an executioner.
- Fluctuating contributions from individual differences are most evident in the case of Activity ratings. According to Table 8.3, no individual differences occur in sensing the liveliness of "murmuring to someone," but the individual variance is 3.67 in assessing the liveliness of an executioner. Constraining all individual variances to be equal yields an average value of 0.93.

Conventional estimates of reliability in measuring each of the 16 concepts in Table 8.3 can be obtained by computing the ratios of individual variances to the sums of individual and error variance. These reliability estimates vary from 0.11 to 0.52, with a median of 0.30 in the case of Evaluation; from 0.08 to 0.62, with a median of 0.32 in the case of Potency; and from 0.00 to 0.62, with a median of 0.20 in the case of Activity. The proportion of individual variance in ratings is less than a third for a majority of concepts, on all three dimensions.

The reliabilities of sentiment measures, with data pooled from all 16 cultural concepts included in the test–retest study, can be calculated from the individual and error variances given in Table 8.3 for the most constrained model (i.e., Equal Individual and Error Variances and Stable Means). The reliability is the individual variance divided by the sum of the individual variance and error variance, as indicated in equation (5.22). Therefore, the average reliability of the Evaluation measures is $0.83/(0.83 + 1.81) = 0.31$. By similar calculations the average reliability for Potency measurements is 0.37, and for Activity the reliability coefficient is 0.30.

8.1.1 Discussion

These reliabilities for measurements of sentiments, amalgamated across cultural concepts, are around half the magnitude of the average reliability of feeling thermometers, as presented in the introduction to this chapter. That raises a question: How can a measuring instrument that is more precise in several respects than a feeling thermometer have much lower reliability?

The incongruity is resolved by considering what is being measured rather than the measuring instruments. The reliabilities of sentiment measures presented above were based on concepts selected to be culturally institutionalized, and as argued in Chapter 1, individuals should be fairly homogeneous in their feelings about cultural concepts. Homogeneity in sentiments about each concept implies that measurement reliabilities must be low, because near-zero individual variance divided by the total rating variance for the concept is a value near zero. In contrast, survey researchers include items in a survey questionnaire only if the items are linked to controversy or social change that promises to generate substantial individual variability. Thus, Alwin's (2007) estimate of the average reliability of feeling thermometers is fairly high because it is based on items selected to have response heterogeneity.

Finding that reliability is low when measuring a cultural sentiment supports the homogeneity argument regarding cultural sentiments. However, the salient question then is: Why are the foregoing estimates of reliability in sentiment measures of individual concepts not zero? Three reasons seem plausible.

First, the assumption that all of the concepts chosen for the test–retest study were culturally institutionalized no doubt was in error for the identity of executioner. Respondents largely agreed in their evaluations of *executioner*, but the large Potency and Activity variances for this concept (in Tables 8.1 and 8.3) show that assessments on these dimensions varied substantially. The high variances probably resulted from two events covered heavily by the mass media prior to the data collection period in the fall of 2003. In January 2003, Governor George H. Ryan of Illinois commuted all death penalties in his state, fueling controversy about capital punishment in the United States. An earlier burst of media attention attended the kidnapping and execution of the American journalist Daniel Pearl by Islamic militants in February 2002. Thus, the cultural meaning of execution and related concepts was in play at the time of the survey.

Chain smoker was another identity with large amounts of individual variance in ratings, and this concept also was culturally in play at the time of the survey. After the turn of the century, tobacco companies were sued successfully for punitive damages, and the World Health Organization developed a tobacco control treaty. However, during the same period the U.S. Supreme Court ruled that advertising that favored smoking could not be banned. Thus, institutionalized denigration of tobacco users had not yet crystallized.

Second, the large individual variance in evaluations of "to wait on someone" (see Table 8.3) might derive from differing interpretations of the stimulus. The item was supposed to be interpreted as the action of an employee in a retail establishment dealing with a customer, and approximately half of the respondents[4] evidently did interpret it this way. However, the stimulus also might have been interpreted as the impatient finger-tapping activity of someone awaiting another, and the negative ratings of the other half of the respondents could reflect this interpretation. Conceivably, some individual variance associated with other stimuli could have arisen similarly from differing interpretations, evoking the normative sentiments of different cultural concepts.

Third, homogeneity of sentiments about a cultural concept is an ideal, whereas sentiment measurements are obtained from real-world respondents who rarely have perfect conformity to norms. Thus, some individual variance in sentiments typically will occur around a cultural norm.

In conclusion, the conventional conception of reliability as the proportion of individual variance in psychological measurements makes little sense when dealing with measurements of cultural sentiments, because homogeneity of individuals within the culture diminishes individual variance in the measurements. The more perfect people's enculturation, the lower the reliabilities of norm measurements when considering one norm at a time.

8.2 RELIABILITIES ACROSS STIMULI

Reliability is the proportion of meaningful variance in a measure, but a different view of meaningful variance is required when assessing cultural sentiments. The meaningful variance comes not from individual differences but from differences in ratings of different concepts.

As seen in equation (5.24), squared individual-to-total correlations, or commonalities, are reliabilities of this type. Thus, the commonalities computed in Chapter 7 can be revisited for some estimates of this type of reliability. Median-squared commonalities among the Business School respondents were 0.68, 0.35, and 0.34 on Evaluation, Potency, and Activity, respectively. The median Evaluation reliability is similar to the average reliability reported by Alwin (2007) for warm–cold feeling thermometers applied to single concepts.

However, measurements of cultural sentiments are based on ratings from all respondents, not just the respondents with median commonality, so it is of interest to develop more comprehensive estimates of reliability. I do this by estimating cultural variance in ratings by all respondents of 1,500 concepts, using analysis of variance procedures, as developed in equations (5.26) through (5.29).

In the overall project, 1,039 respondents in 15 subsamples each rated 100 concepts—a total of 1,500 different concepts. To assess the amount of variance in ratings that is associated with cultural norms, I conducted 45 analyses of variance of these data, separately for the Evaluation, Potency, and Activity

[4]The distribution of ratings has two major modes, one at +1.0 and another at −1.0. K-means cluster analysis of the data, with specification of two groups, results in estimation of one group's mean at +1.77 and the other group's mean at −1.52.

measurements in each of the 15 subsets of data involving different raters and different concepts. Each two-way analysis of variance yielded three statistics of interest: the variance of the means for the 100 concepts, the variance in the means of respondents across all 100 concepts which I call respondent bias), and the residual variance. I employed a mixed-linear model (Norušis 2003, Chapter 23) to obtain a (restricted) maximum-likelihood solution in the presence of variable numbers of raters for different concepts. The top half of Table 8.4 summarizes the results of these analyses.

TABLE 8.4 Components of Variance in Ratings of 1,500 Concepts Distributed into 15 Stimuli Sets

	Cncpt Var	IB Var	Res Var	Propor Cncpt
		All Respondents		
		Evaluation		
Minimum	2.29	0.05	2.10	.450
Maximum	4.00	0.16	3.00	.613
Median	2.88	0.08	2.70	.514
		Potency		
Minimum	0.92	0.14	2.81	.194
Maximum	1.50	0.35	3.54	.321
Median	1.15	0.20	3.08	.257
		Activity		
Minimum	0.80	0.08	2.79	.191
Maximum	1.53	0.25	3.49	.328
Median	1.07	0.14	3.04	.251
		High-Commonality Respondents		
		Evaluation		
Minimum	4.07	0.03	1.89	.620
Maximum	6.25	0.11	2.65	.748
Median	4.68	0.07	2.20	.668
		Potency		
Minimum	1.66	0.04	2.28	.341
Maximum	2.64	0.12	3.57	.448
Median	2.11	0.08	3.12	.400
		Activity		
Minimum	1.47	0.02	2.49	.316
Maximum	2.84	0.09	3.77	.475
Median	1.82	0.05	3.09	.359

Note. Abbreviations: Concept Variance: Cncpt Var; Individual Bias Variance: IB Var.; Residual Variance: Res Var; Proportion of Total Variance for Concepts: Propor Cncpt.

On the Evaluation dimension, concept variations explained from 45 to 61 percent of the total variance in ratings over the 15 datasets, and the median was 51 percent. Variance associated with respondent biases was small but highly significant in all cases, ranging from 1 to 2.5 percent, with a median of 1.4 percent. Residual variance in these analyses combines individual variations in rating different concepts and measurement-error variance, and the median value of 2.70 was close to the value of 2.64 obtained in the test–retest analyses of 16 concepts (i.e., 2.64 is the sum of 1.81 and 0.83, shown at the bottom of Table 8.3).

On the Potency dimension, concept variations explained from 19 to 32 percent of the rating variance, with a median of 26 percent. Variance due to respondent biases was larger here, ranging from 3 to 7 percent, with a median of 4.6 percent. The median residual value of 3.08 again was similar to the estimate obtained via test–retest analyses and given in Table 8.3: 1.91 + 1.12 = 3.03.

On the Activity dimension, from 19 to 33 percent of the total variance was explained by concept means, with a median of 25 percent. Variance due to respondent biases ranged from 2 to 6 percent, with a median of 3.3 percent. The median value of residual variance, 3.04, once again was close to the alternative estimate given in Table 8.3: 2.22 + 0.93 = 3.15.

As noted in the discussion following equation (5.28), analyses based on all respondents, including some respondents with less than perfect cultural inculcation, underestimate the proportion of rating variance associated with cultural norms among well-inculcated individuals. To appreciate the levels of variance explained by cultural norms among well-inculcated individuals, I recomputed the 45 analyses of variance, employing only raters with above-average cultural inculcation, as indicated by their having an inculcation score above the median. The results are reported in the bottom half of Table 8.4.

Variance associated with concepts increased substantially when computing concept means only from ratings of respondents who are culturally well inculcated. The small amounts of variance due to respondents' biases were reduced,[5] and estimates of variance due to idiosyncratic and measurement-error variations stayed approximately the same. The net effect is that cultural norms explain more rating variance for well-inculcated respondents. Specifically, the median amount of variance explained by concept means increases from 51 percent to 67 percent on the Evaluation dimension; from 26 percent to 40 percent on the Potency dimension, and from 25 percent to 36 percent on the Activity dimension. As proportions, these figures constitute reliability coeffi-

[5]Respondent bias is the difference between a person's average rating over all concepts and the average rating computed over all individuals. Cultural norms cancel out of this calculation if everyone has perfect cultural inculcation. However, equation (15) shows that the averaged cultural norm is a factor when some individuals have imperfect inculcation and averaged norms are not zero. Estimating averaged cultural norms from all ratings of all 1,500 concepts reveals nonzero means: 0.29 on Evaluation, 0.64 on Potency, and 0.48 on Activity, Thus, the apparent biases are greater in the sample of respondents with mixed levels of inculcation than in the sample of well-inculcated respondents.

cients for sentiment measures when respondents with above-average inculcation of cultural sentiments rate a typical range of cultural concepts.

8.2.1 Reliabilities of Means

In practice, ratings from multiple respondents are averaged in order to estimate cultural sentiments, and the standard error of a mean is smaller than the standard deviation of the individual ratings. Therefore, a mean reduces error variance in estimating an average value, or norm. In particular, a mean computed over N respondents has a variance that is $1/N$ the size of the variance of individual ratings. Because the variance of mean ratings contains a smaller proportion of irrelevant variance and a correspondingly larger proportion of meaningful variance, the reliabilities of mean ratings increase as larger numbers of respondents are used to compute means.

Both measurement errors and individual differences are sources of inaccuracy when trying to estimate cultural norms in sentiments. Thus measurement-error variance and individual variance need to be summed to define the respondent-error variance that arises when assessing cultural sentiments. I use the variance estimates for individual differences and error obtained with the most constrained model presented in Table 8.3. Table 8.5 gives the standard deviations due to individual differences and measurement errors when means are computed over various numbers of respondents.

Standard deviations of 1.6 on Evaluation, 1.7 on Potency, and 1.8 on Activity are reasonable general estimates of error when a single respondent is being employed to assess cultural sentiments. Averaging ratings over as few as five respondents approximately halves the standard deviations of sentiment measures. Averaging over 20 respondents makes measurements of normative sentiments about four times more accurate than ratings from a single respondent. Averaging over 50 respondents makes the measurements about seven times more accurate, reducing the standard deviation to about 0.25 on a scale with a range of 9.6.

TABLE 8.5 Standard Errors of Mean Sentiment Ratings for Different Sample Sizes

Sample Size	Evaluation	Potency	Activity
1	1.62	1.74	1.77
5	0.73	0.78	0.79
10	0.51	0.55	0.56
15	0.42	0.45	0.46
20	0.36	0.39	0.40
30	0.30	0.32	0.32
40	0.26	0.28	0.28
50	0.23	0.25	0.25

Note. Each standard error is computed as SQRT(S2/N), where S2 is the sum of individual and error variance in the bottom row of Table 8.3.

TABLE 8.6 Reliabilities of Mean Sentiment Measures, Based on Average Standard Errors and Median Variances of Means for 1,500 Concepts

Sample Size	Evaluation	Potency	Activity
1	0.52	0.28	0.25
5	0.84	0.65	0.62
10	0.91	0.78	0.76
15	0.94	0.85	0.83
20	0.96	0.88	0.87
30	0.97	0.92	0.91
40	0.98	0.94	0.93
50	0.98	0.95	0.94

Note. Table 8.4 gives the median values of variances due to concepts in 15 sets of 100 concepts: 2.88, 1.15, and 1.07, for Evaluation, Potency, and Activity, respectively. Each reliability is calculated as a concept variance divided by the sum of the concept variance and the appropriate squared standard error of the mean from Table 8.5.

Table 8.6 presents the results of computing the reliabilities of sentiment measurements from the average concept variances in Table 8.3 and from standard errors of mean ratings for the array of sample sizes given in Table 8.5. Table 8.6 shows that reliabilities decline in going from Evaluation measurements to measurements of Potency and Activity, which is a combined result of greater errors and individual differences associated with evaluative reduplication in measuring Potency and Activity norms and of smaller amounts of cultural variation on the Potency and Activity dimensions. Nevertheless, reliabilities above 0.90 on all three dimensions of sentiment measurement are obtained when ratings are obtained from 30 or more respondents. In fact, just 26 respondents are required to meet this criterion.

To check that reliabilities of the normative measures are of this order, I correlated female and male means over the 1,500 concepts considered in this study, where the median number of female raters was 34 and the median number of male raters was 32. The female–male correlations were very high, notwithstanding the substantive gender differences documented in Table 7.5: 0.96 on Evaluation, 0.89 on Potency, and 0.89 on Activity. Such high gender correlations could not be attained unless reliabilities of the normative measures were even higher[6] such as the values indicated for samples of 30 raters in Table 8.6.

The median proportions of variance due to concepts in ratings of 1,500 concepts, as given in the top half of Table 8.4, offer alternative estimates of reliabilities. These estimates are close to the more sophisticated reliability estimates given in the first line of Table 8.6. The bottom half of Table 8.4 gives

[6]The reliability of female measurements of sentiment norms is slightly higher than the reliability of male measurements because of females' somewhat greater enculturation in the affective realm (see Table 8.2).

median proportions of concept variance for respondents with above-median commonalities (i.e., reliability estimates for "expert" respondents) of 0.67 on Evaluation, 0.40 on Potency, and 0.36 on Activity. Converting these figures to reliabilities of mean ratings using a standard formula in measurement theory (Nunnally 1967, Equation 6-18) reveals that reliabilities of 0.90 or more are achieved with just 16 above-median respondents, as opposed to 26 unscreened respondents. Although there might be no gain to restricting analyses to above-median respondents after data have been collected, one could collect data from about a third fewer respondents than usual if expertise could be ascertained before data collection begins, and data collected from expert respondents only.

Winkler and Hays (1970, Equation 11.10.4) presented a formula that uses analysis of variance results to obtain the proportion of variance in a dependent variable that is accounted for by experimental treatments:

$$\frac{SS_{\text{between}} - (J-1)\cdot MS_{\text{within}}}{SS_{\text{total}} + MS_{\text{within}}} \tag{8.1}$$

SS stands for sum of squares, *MS* for mean square, and *J* is the number of treatments. To assess how this serves for obtaining estimates of reliabilities of sentiment measurements, I computed one-way analyses of variance of (ANOVA) EPA ratings for the 1,500 concepts in this book's survey, male and female ratings pooled. Substituting ANOVA results for Evaluation into formula (3.1) gave the following:

$$\frac{304085.833-(1499\cdot 2.752)}{574183.317+2.752} = 0.52 \tag{8.2}$$

The reliability estimated this way is the same as that reported in the first row of Table 8.6. Parallel computations for Potency and Activity gave a value of 0.26 in both cases. These values are close to the Potency and Activity reliabilities given in the first row of Table 8.6.

Of course, the estimates of reliability provided in this section can change in different domains of stimuli, since the amount of meaningful variance in cultural sentiments depends on which concepts are being studied. To illustrate, I used the Winkler–Hays formula to compute separate reliabilities for the 500 identities, 500 behaviors, 200 settings, and 300 modifiers in this book's survey. The results were as follows, with numbers representing Evaluation, Potency, and Activity reliabilities, respectively:

- *Identities:* 0.48, 0.29, 0.21
- *Behaviors:* 0.54, 0.17, 0.16
- *Settings:* 0.37, 0.16, 0.33
- *Modifiers:* 0.59, 0.34, 0.33

The figures suggest that more respondents are needed to build good measurements of behavior and setting potencies, of behavior activities, and of setting evaluations; and that modifier sentiments can be estimated with somewhat fewer respondents than are needed for the other types. However, these variations in reliabilities correlate 0.94 with the variances of the corresponding variables. For example, modifier evaluations had the largest variance (6.524), while behavior activities had the smallest variance (3.669). Thus, the differences in reliabilities mostly reflect differences in meaningful variance for each concept type relative to fairly constant error variances on each dimension.

The amount of meaningful variance in a domain of concepts can also vary from one culture to another; for example, variability in evaluations of nuclear family identities is much greater in Japan than in the United States (Heise 2007, Figure 3.1).

8.2.2 Discussion

A reliability coefficient reports the proportion of meaningful variance within the total variance of measurements, where the total variance consists of both meaningful variance and error variance. When measuring cultural norms, there is relatively little individual variance in sentiments about a cultural concept, so traditional reliability coefficients that treat individual variance as meaningful tend to have low values. Instead, one must treat variance in sentiments across concepts as the meaningful component and aggregate individual variations in sentiments with measurement errors, since individual variance interferes with inferring cultural norms. Such understandings led to a set of reliability estimates for sentiment ratings when employed as measures of cultural norms. The reliabilities of a typical respondent's ratings as a measure of norms turned out to be modest with this approach, although the reliability on the Evaluation dimension was comparable to the reliabilities of feeling thermometers as reported by Alwin (2007, Table 9.3). However, the reliabilities of mean ratings computed over a group of 30 or more respondents were above 0.90 on all three dimensions, revealing that cultural sentiments can be measured reliably by averaging ratings of multiple respondents.

Throughout the world of science, a standard method of reducing error variance is to compute the mean of multiple measurements. Table 8.6 shows the dramatic effects that this has on reliabilities in culture studies. Whereas one person's indication of cultural sentiments may be of marginal value, the mean ratings of 30 people provide high-quality estimates of cultural norms.

8.3 CHAPTER HIGHLIGHTS

- Individual variance was a third or less of the total variance in rating sentiments for a majority of the 16 concepts in the test–retest study. Individual variance exceeded 50 percent of the total rating variance only for concepts where cultural change was occurring (and for a concept whose meaning was ambiguous to raters). In consequence, the reliability of sentiment measurements typically was low for these concepts, when reliability was conceived as the ratio of individual variance to total rating variance.

- Finding low conventional reliabilities for sentiment measures supports the homogeneity assumption regarding cultural sentiments: that is, that respondents' personal sentiments largely reflect cultural norms, so variations in ratings mostly reflect measurement errors rather than individual differences in sentiments.
- The reliabilities of concept Evaluations were comparable to the reported reliabilities of feeling thermometers, when the reliability of Evaluations was conceived as the proportion of total variance associated with variations in concepts. Potency and Activity reliabilities were about half as large because Potency and Activity ratings are subject to more measurement errors and to reduplication of Evaluations by some respondents, and also because concept meanings have less variation on Potency and Activity than on Evaluation.
- Averaging ratings from multiple respondents increases the reliability of sentiment-norm measurements. Reliabilities above 0.90 on all three dimensions of affective response can be achieved by averaging ratings of 30 respondents. Fewer respondents are needed for this level of precision if all of the respondents are preselected for high levels of enculturation.

9 Culture and Surveys

Surveys of culture have roots in traditional ethnography on the one hand and conventional survey research on the other. Notwithstanding these antecedents, however, surveys of culture have distinctive features that differentiate them from both traditional ethnographic studies and from conventional survey research studies.

9.1 UNIQUE ASPECTS OF SENTIMENT SURVEYS

Traditionally, ethnographers have delineated cultures by questioning several well-inculcated informants and generalizing the information that they provide to the culture as a whole. Could such an approach provide sentiment repositories like those that were crucial in developing affect control theory (Heise 2007)? To examine the issue in more detail, I imitated the traditional ethnographic approach in a small way by rating the 575 stimuli given in the appendix to Chapter 4. To obtain information on reliability, I rated each stimulus three times, in 24 sessions spread over two months, devoting somewhat over 10 hours total to the task. Below, I speak of my ratings as those of an informant—one who is well-inculcated into American middle-class culture, conscientious, and motivated enough to labor through the hundreds of ratings on computer-presented scales.

The reliabilities of the informant's single ratings, computed with equation (8.1), were 0.92, 0.83, and 0.83 on Evaluation, Potency, and Activity, respectively. The reliabilities of the informant's mean ratings of concepts, averaged over three occasions, were 0.97, 0.94, and 0.94. Corresponding reliabilities of single ratings by a typical respondent in the survey employed throughout this book, as reported in Table 8.6, were 0.52, 0.26, and 0.26; and the same table reports that the reliabilities of means based on 30 respondents would be about 0.97, 0.92, and 0.91. Thus averaging three ratings by one adept, well-motivated informant gives levels of precision that correspond to averages of 30 ratings by ordinary survey respondents.

While mean ratings from the informant were precise measurements of the informant's sentiments, the informant's sentiments were not equivalent to

---

cultural sentiments because the informant's sentiments were determined by personal experiences as well as by cultural norms. To assess how close the informant's sentiments were to cultural sentiments, I correlated the informant's mean ratings of 109 identities, behaviors, and modifiers with mean ratings of the same concepts by about 30 males in the survey reported in Chapter 3. The correlations of 0.93, 0.81, and 0.85 are comparable to correlations in sentiments between the United States and Canada (Heise 2001a, Table 2). Thus, employing the informant as the sole source of information about sentiment norms in the United States would be like doing a sentiment study in Canada in order to describe American culture. Such an expedient generally would be irrational (except, perhaps, if the United States were an unknown culture and the informant provided the only practical avenue for learning about that culture).

As opposed to traditional ethnographic studies, numerous respondents must participate in defining cultural norms in the survey-of-culture approach for three reasons. First, one cannot count on recruiting high-quality respondents such as the informant considered above, especially considering the time demands on such a respondent to rate and re-rate stimuli. One must presume that average respondents are recruited in a survey, of the kind considered in Chapter 8, and a score or more of such respondents is required to attain suitable measurement reliabilities. Second, as revealed in the foregoing analysis of a single informant, multiple respondents are needed to swamp the idiosyncratic components of each respondent's personal sentiments. Third, no individual respondent can be expected to provide data for the many concepts of interest in a cultural study—the task needs to be distributed across many respondents. For example, in the sentiment survey examined in this book, the challenge of dealing with 1,500 concepts was met by randomly sampling from a respondent pool of hundreds of people in order to get comparable samples of respondents to rate subsets of concepts.

As indicated above, surveys of culture differ from traditional ethnographic methods. Surveys of culture also differ from traditional forms of survey research in several ways. Expositions on cross-cultural surveys, such as the book of Harkness, van de Vijver, and Mohler (2002), can provide valuable guidance for investigators undertaking culture surveys. Both kinds of work must deal with such issues as translation, assessment of respondents' background variables, nonresponse, and documentation of procedures, so researchers doing culture surveys can benefit from the substantial wisdom accumulated by researchers doing cross-cultural surveys. Nevertheless, several features demarcate surveys of culture from cross-cultural surveys.

- Surveys of culture generally focus on a single culture and take an emic, culturally deferential orientation, albeit with possible preparations for surveys of additional cultures so that research can culminate in comparative studies. In contrast, cross-cultural surveys deal mainly with etic, culturally neutral variables that can be assessed equally well throughout a given sample of societies, notwithstanding the cultural variations in those societies.

- Surveys of culture aim at discovering norms—shared conceptions and sentiments. In contrast, cross-cultural surveys focus on comparing populations' central tendencies and dispersions, along variables preselected to register substantial individual differences.
- In surveys of culture, sampling means acquiring respondents who are most authoritative regarding their culture, and such experts typically are found at specific locales where the culture is being constructed and reproduced. Sampling in cross-cultural surveys means acquiring respondents—wherever they may be—who as a group represent the societal populations from which they were drawn. In the words of Häder and Gabler (2003, p. 117), in cross-cultural research, deviations from textbook norms of sampling "are acceptable departures provided probability samples and only probability samples are used."

Because the differences between surveys of culture and cross-cultural surveys are not widely appreciated, critics frequently feel quite certain that surveys of culture do not meet widely accepted standards of survey research. A brief aside drawn from the history of survey research offers some perspective.

Converse (1987, p. 85) noted that in the 1920s and 1930s academic social scientists designed instruments of considerable elegance to quantify attitudes, and they inaugurated ground-breaking investigations of how attitudes relate to other matters. "Their work was nevertheless limited by its sampling. Even the range of variation in attitude available to study was circumscribed by the use of student groups, and when researchers broke out of academic environments, they put together groups from the adult population in the patchworks we have observed. That limitation was linked to another: that of organization. The social psychologists proliferated 'small' science, projects directed by one or two individuals, assisted at best by a few graduate students. The work necessarily lacked scope and continuity because academic researchers simply did not have the necessary money and institutional capacity."

Converse's criticisms of the pioneering early social psychologists have a familiar ring for contemporary researchers involved in surveys of culture. For example, journal reviewers frequently complain that sentiment ratings by university students do not constitute a probability sample and therefore cannot be representative of the national population. I have argued in this book that complaints against nonprobability sampling in culture surveys are misdirected. Studies of norms do not seek a range of cultural influences among respondents, but just the opposite—homogeneity, and all respondents are not equally adequate for the task of assessing cultural norms.

For example, it is arguable that the homogeneity of university students is advantageous in defining norms efficiently and that students are especially germane to the task of cataloging middle-class sentiment norms. The relevant sampling problem is not dependence on students but the fact that most sentiment studies so far have depended only on students in single-cluster sampling. Even the multicluster exceptions, such as the study considered in this book, drew geographically dispersed locations from within the single institution of education. A proper study of middle-class sentiment norms would employ multicluster sampling across multiple social institutions in order to acquire respondents who are proficient in reproducing multiple aspects of middle-class culture.

This example illustrates that a criticism is warranted regarding survey-of-cultures samples, but it is not the one usually made. Rather, the problem is that theory and practical procedures for sampling the settings of culture production and reproduction still are undeveloped. In fact, the development of this topic is about at the stage where the theory and methods of sample surveys of people were early in the twentieth century (Converse 1987, pp. 39–49).

Converse's criticism of the pioneer social psychologists engaging in "small" science also pertains to many survey-of-culture projects. For example, each of the sentiment repositories in six nations and four languages discussed in Chapter 3 was compiled under the direction of one or two principal investigators, usually with the help of graduate students. Needs for funding were modest because (1) questionnaires were group administered in early studies and self-administered in later Internet studies; (2) ratings made on graphic rating scales were recorded directly in quantitative form without human coding; (3) technologies of optical mark recognition in early studies and Java applets in later studies were adopted to eliminate manual keying of respondent's answers; and (4) a standardized format for background questions and for presentation of rating stimuli eliminated the high cost of retooling for each new survey, a particularly costly issue in computerized questionnaires (Schonlau, Fricker, and Elliot 2002, p. 78). The economy of these studies made explorations and discoveries possible in the absence of large-scale funding such as that required for sample surveys.

However, Converse's point about the disadvantages of noninstitutionalization are valid. No organizational mechanism exists to foster creation of sentiment repositories in new languages and cultures or to replicate repositories in order to study cultural change. Expansion and replications have occurred, but entirely as a product of the energy and initiative of a few research entrepreneurs.[1]

Some have suggested that surveys of cultural sentiments might be piggybacked within standard sample surveys. Doing so conceivably might mitigate the institutionalization problem by eventually adding a cultural studies unit to the organization chart of some survey research institutes. It also might enable

[1]Herman Smith, Lynn Smith-Lovin, Andreas Schneider, and Neil MacKinnon each organized assembly of two repositories. I have done five, two in cooperation with Smith-Lovin.

the identification of distinct complexes of sentiment norms corresponding to different cultures or subcultures within a national population. However, this development is not likely, for the following reasons.

First, incorporating sentiment surveys into sample surveys would not apply the right sampling design for sentiment surveys. Sample surveys are designed to get representative samples of individuals within some political boundaries, whereas sentiment surveys ideally must sample settings where culture is being constructed and reproduced by culturally homogeneous individuals. Second, inclusion of sentiment studies in a sample survey would require auxiliary questions being included to ascertain what culture the respondent might be in. Third, respondents would have to be trained to use the rating instrument. Scales for measuring sentiments are comparable to feeling thermometers in survey research. However, three scales, or thermometers—one each for E, P, and A—must be presented for each stimulus in order to measure sentiments, and use of the three scales would require some instruction. The computerized instruments for assessing sentiments do not take long for educated respondents to learn—5 minutes or so—but even a small amount of training time could preclude use within the tight time constraints of sample surveys. Fourth, the full power of computerized rating instruments could be obtained only in Web-based surveys or during face-to-face interviews. Accurate sentiment measurement probably could not be attained at all with the kinds of questions asked in telephone surveys. Finally, only a few stimuli for sentiment ratings could be included in a particular form of a sample survey. Say that each respondent rates six stimuli on three dimensions, that 1,000 or more stimuli are required for a sentiment repository, and that 30 respondents must rate each stimulus in order to achieve adequate precision; then simple calculations reveal that a study similar to the one analyzed in this book could be achieved only in sample surveys with more than 150 questionnaire variations and 5,000 or more respondents. Sample surveys of this magnitude are atypical.

Notwithstanding all of these problems, piggybacking can work in some instances. The Rossi and Rossi (1990) study of kinship norms is a prime example of piggybacking a norm study on a sample survey. That particular case worked well because respondents were drawn from households in Boston, yielding a fairly homogeneous group of raters with regard to family norms, because the task of judging obligation was simpler than measuring sentiments on three dimensions, because the conveying sample survey involved face-to-face interviews with a large sample of respondents, because randomly sampling stimuli (vignettes) made sense in the study, and because respondents were interested enough in the material to complete more than a few ratings.

9.2 FRAMEWORKS FOR SENTIMENT SURVEYS

Consider two conceivable frameworks for sentiment surveys: an ethnological framework which presumes that cultural norms are the key source of variation in ratings, and an individual-differences model that presumes that disparities among respondents are the key source of variation in ratings. An ethnological model assumes that respondents who are well inculcated into a culture give the same normative responses, except for errors of measurement. Respondents' ratings of a particular concept should correlate very little over time because the rating variations are mostly measurement errors, nonpredictable from one time to another. However, ratings should be substantially different across concepts, to the extent that cultural norms set different sentiments for the different concepts.

The individual-differences model assumes individuality of responses, grounded in individuals' unique experiences, with measurements from different individuals being independent. Respondents' ratings of a particular concept generally should correlate over time as the respondents repeatedly report their idiosyncratic sentiments. However, mean ratings for different concepts should tend to regress toward an overall mean, since every concept would have a mix of respondents with positive and negative sentiments. The results obtained in Chapter 8 regarding sentient measurements align with both frameworks to a degree.

I found over-time correlations in sentiment ratings of about 0.35 on the average, indicating that about one-third of the variance in rating a single concept comes from stable individual differences in sentiments. I used structural equation modeling and analyses of variance to obtain more refined estimates of individual variance and found about the same proportion of meaningful individual variance in ratings of a single concept. I then used analyses of variance and structural equation modeling to partition the overall variance obtained when rating diverse concepts. The results varied for the three dimensions of Evaluation, Potency, and Activity.

On the Evaluation dimension, cultural norms accounted for about half of the overall rating variance (and this figure jumped to two-thirds when respondents were restricted just to those who had high commonality with others). Since a third of the other half of the variance came from stable individual differences, individual differences accounted for one sixth of the overall rating variance. The rest—measurement errors—amounted to one-third of the overall rating variance.

In the case of both Potency and Activity, cultural norms determined about one-quarter of the total rating variance (the proportion was higher for high-commonality respondents: 40 percent in the case of Potency and 35 percent on Activity). With a third of the remaining variance coming from stable individual differences, individual differences also accounted for one-quarter of the overall rating variance. Measurement errors constituted half of the total rating variance.

Cultural norms determined a larger proportion of variance on the Evaluation dimension than on the Potency and Activity dimensions, in part because Evaluation norms were more extreme. That is, concepts with highly polarized mean ratings were more common on the Evaluation dimension than on the Potency and Activity dimensions. Additionally, individual differences constituted a larger proportion of the total variance on the Potency and Activity dimensions than on the Evaluation dimension because some respondents reduplicated Evaluation norms in their Potency or Activity ratings, as discussed in Chapter 7.

Individual variance in ratings departs from the ethnological framework's ideal of respondents whose sentiments are culturally homogeneous. Variance in ratings due to cultural norms departs from the individual-differences framework's ideal of respondents whose sentiments are so independent and variable that differences in means for different concepts almost disappear. On the Evaluation dimension, the balance favors the ethnological framework, with one-sixth of the variance relating to individual differences and half of the variance being normative. On the Potency and Activity dimensions the two sources of variance are equal. Overall, the ethnological framework provides a credible framework for the interpretation of sentiment ratings regarding stable cultural constructs, although an individual-differences framework is equally credible for Potency and Activity measurements.

The appropriateness of an ethnological framework for sentiment studies was reiterated by the Q-factoring results in Chapter 7. A single factor massively dominated Evaluation ratings, revealing that respondents mostly agreed in their feelings about the goodness or badness of different concepts, aside from individuals' varying propensities to diminish or intensify normative feelings on rating scales. Dominant factors were also found in Potency and Activity ratings, indicating that sentiments on these dimensions also were normative, although a substantial second factor uncovered the fact that individuals had varying propensities to reduplicate evaluation norms in their Potency or Activity ratings.

In the ethnological framework for studying cultural norms, individual variations in sentiments turn into measurement errors; that is, people's deviations from normative sentiments interfere with identifying norms. Nevertheless, the individual variance may be of substantive interest. For example, Kroska and Harkness (2006) found that stigmatization of the mentally ill correlated positively with individual variations in sentiments about identities such as loser and reject and correlated negatively with sentiments about identities such as doctor and hero, even though ratings of all the identities were controlled primarily by cultural norms. On the other hand, analyses in this book suggest a possible new interpretation for such findings: Correlations involving individual differences in sentiments may arise from different levels of cultural inculcation among respondents. Thus, the pattern of correlations that Kroska and Harkness found could have arisen because the stigmatizers were better-inculcated respondents who exalted the normatively valued and degraded the normatively disvalued, including the mentally ill. The formal analysis of cultural inculcation and sentiment measurements in Chapter 5 indicated that respondent variations in levels of cultural inculcation can engender covariances among items measuring normative sentiments that suggest artifactual congeries of items involving the most extreme sentiments.

9.3 IN CLOSING

Paul Lazarsfeld (1972, p. 290) stated a motif of this book: "Distinguished equally by brilliance and by irresponsibility of factual evidence, they [cultural anthropologists] challenge the pollster to try to bring about effective cooperation. But the challenge is worth accepting, for from an interaction between the two groups could develop really new insights into human affairs." In fact, ethnographers have not been irresponsible in their data gathering; an ethnographer's field notes may contain more items of information than a survey does. However, I believe that Lazarsfeld was correct in the rest of his statement. New ways of understanding culture and society result from merging ethnography and survey research into a distinctive research framework. Whatever technical difficulties may be involved in such an endeavor, a major resistance appears to be disciplinary.

On the ethnographers' side, a double lambasting from postmodern critiques, followed by global homogenization of cultures, has fueled a contemporary retreat from describing enduring systems of cultural meanings (Barrett 2002, p. 1) and sometimes generated a denial of the existence of such systems altogether (even though the denial must be stated in terms of enduring linguistic meanings). Apart from that, the spirit of enterprising qualitative ethnographers listening attentively to people lives on, so there is little sufferance for notions of nonproximate relations with informants. The instrumentation used in a survey of culture, such as the computer-implemented rating scales discussed in this book, may be foreign and suspect for traditional ethnologists.

The monumental accomplishments of survey research over the last half century instilled great—even overweening—confidence in this research approach. One major achievement was the technique of probability sampling, which allows relatively small samples of respondents to represent vast populations of individuals. Probability sampling is now so consecrated among survey researchers that suspicion might meet the argument in this book that this technique is an inappropriate way of choosing respondents for culture studies. Another accomplishment in survey research has been development of an array of methods for evaluating questionnaire items, including statistical measures of reliability. Accordingly, the argument here that an ideal measure of norms has zero reliability in the traditional sense may be grating, even seeming ridiculous, to survey researchers.

I have tried to address such discomforts in this book, especially those of survey researchers and quantitative methodologists. [Another disquisition (Heise 2007, Part I) attempts to show the interpretive value of sentiment surveys.] The measure of my success will be the extent to which problems of culture surveys are set upon in the future in order to produce another branch of survey research as successful as the first.

9.4 CHAPTER HIGHLIGHTS

- A well-inculcated and well-motivated respondent can produce high-quality information on cultural norms. However, the information obtained is biased toward that respondent's personal sentiments. Moreover, the respondent rarely would be willing to commit enough time to rate hundreds of concepts. Thus, studies of cultural sentiment norms inevitably depend on samples of respondents rather than on a few expert informants.
- Surveys of culture are a different form of research than cross-cultural surveys. The two are distinguished in several ways: an emic approach as opposed to an etic approach, a focus on norms versus a focus on individual differences, and disparate approaches to sampling respondents.
- Surveys of cultural sentiments can only rarely be conducted within the context of standard survey studies. A survey study ordinarily would impose the wrong sampling scheme, and typical survey studies have little room for the extra instructions and many questions involved in a study of cultural sentiments. Telephone surveys in particular are unsuitable for surveying cultural sentiments.
- On the whole, surveys of sentiments associated with diverse concepts fit an ethnological framework better than an individual-differences framework. An ethnological framework presumes that cultural norms are the key source of variation in ratings, whereas an individual-differences framework presumes that disparities among respondents are the key source of variation in ratings. An ethnological framework is particularly apropos for evaluative ratings—a finding that was confirmed in several different ways in this book.

References

Alves, Wayne M. 1982. "Modeling distributive justice judgments." Pp. 205–234 in *Measuring Social Judgments: The Factorial Survey Approach*, edited by Peter H. Rossi and Steven L. Nock. Beverly Hills, CA: Sage Publications.

Alwin, Duane F. 2007. *Margins of Error: A Study of Reliability in Survey Measurement*. Hoboken, NJ: Wiley-Interscience.

Anderson, Norman H. 2008. *Unified Social Cognition*. New York: Psychology Press.

Arbuckle, James L., and Werner Wothke. 1999. *Amos 4.0 User's Guide*. Chicago: SPSS Inc. and SmallWaters Corp.

Averett, Christine P. 1981. "Attribution of interpersonal qualities: an affect control theory approach" (Ph.D dissentation). Chapel Hill, NC: University of North Carolina.

Averett, Christine P., and David R. Heise. 1987. "Modified social identities: amalgamations, attributions, and emotions." *Journal of Mathematical Sociology* 13:103–132.

Azar, Edward E., and Steven J. Lerner. 1981. "The use of semantic dimensions in the scaling of international events." *International Interactions* 7:361–378.

Barrett, Stanley R. 2002. *Culture Meets Power*. Westport CT: Praeger.

Batchelder, William H. 2009. *Cultural Consensus Theory: Aggregating Expert Judgments about Ties in a Social Network*. Irvine, CA: Department of Cognitive Sciences, University of California.

Batchelder, William H., Ece Kumbasar, and John P. Boyd. 1997. "Consensus analysis of three-way social network data." *Journal of Mathematical Sociology* 22:29–58.

Batchelder, William H., and A. Kimball Romney. 1988. "Test theory without an answer key." *Psychometrika* 53:71–92.

Berk, Richard A., and Peter H. Rossi. 1977. *Prison Reform and State Elites*. Cambridge, MA: Ballinger.

———. 1982. "Prison reform and state elites: a retrospective." Pp. 145–175 in *Measuring Social Judgments: The Factorial Survey Approach*, edited by Peter H. Rossi and Steven L. Nock. Beverly Hills, CA: Sage Publications.

Bhatt, Rakesh M. 2001. "World Englishes." *Annual Review of Anthropology* 30:527–550.

Biemer, Paul P., Robert M. Groves, Lars E. Lyberg, Nancy A. Mathiowetz, and Seymour Sudman (Eds.). 2004. *Measurement Errors in Surveys*. Hoboken, NJ: Wiley-Interscience.

Bogomolny, A. 2008. *Infinite Latin Squares, from Interactive Mathematics: Miscellany and Puzzles*. www.cut-the-knot.org/arithmetic/infiniteLatin.shtml

Bradley, Margaret M., and Peter J. Lang. 1994. "Measuring emotion: the self-assessment manikin and the semantic differential." *Journal of Behavioral Therapy and Experimental Psychiatry* 25:49–59.

———. 1999. *Affective Norms for English Words (ANEW): Stimuli, Instruction Manual and Affective Ratings*. Gainesville, FL: Center for Research in Psychophysiology, University of Florida.

Britt, Lory, and David R. Heise. 1992. "Impressions of self-directed action." *Social Psychology Quarterly* 55:335–350.

Carlson, David. 2001. "Notes from the city: an urban fieldworker's experience." *DARE Newsletter* 4:1–3.

Carver, Craig M. 1985. "The *DARE* map and regional labels." Pp. xxiii–xxxv in *Dictionary of American Regional English*, edited by Frederic G. Cassidy. Cambridge, MA: Belknap Press.

Cassidy, Frederic G. (Ed.). 1985. *Dictionary of American Regional English: Introduction and A–C*. Cambridge, MA: Belknap Press.

Chapin, F. Stuart. 1932. "Socio-economic status: some preliminary results of measurement." *American Journal of Sociology* 37:581–587.

Chave, E. J., and L. L. Thurstone. 1929. *The Measurement of Attitude: A Psychophysical Method and Some Experiments with a Scale for Measuring Attitude Toward the Church*. Chicago: University of Chicago Press.

Clore, Gerald L., and Jesse Pappas. 2007. "The affective regulation of social interaction." *Social Psychology Quarterly* 70:333–339.

Cohen, Albert. 1955. *Delinquent Boys*. New York: Free Press.

Converse, Jean M. 1987. *Survey Research in the United States: Roots and Emergence 1890–1960*. Berkeley, CA: University of California Press.

Cross, Susan E., and Laura Madson. 1997. "Models of the self: self-construals and gender." *Psychological Bulletin* 122:5–37.

D'Andrade, Roy G. 1987. "Modal responses and cultural expertise." *American Behavioral Scientist* 31:194–202.

———. 1995. *The Development of Cognitive Anthropology*. New York: Cambridge University Press.

Denzin, Norman K., and Yvonna S. Lincoln (Eds.). 2005. *The SAGE Handbook of Qualitative Research*. Thousand Oaks, CA: Sage Publications.

Ekman, Paul. 1971. "Universals and cultural differences in facial expressions of emotion." In *Nebraska Symposium on Motivation: 1971*, edited by J. K. Cole. Lincoln, NE: University of Nebraska Press.

———. 2004. *Emotions Revealed: Recognizing Faces and Feelings to Improve Communication and Emotional Life*. New York: Henry Holt & Company.

Elliott, Lois L., and Percy H. Tannenbaum. 1963. "Factor structure of semantic differential responses to visual forms and prediction of factor-scores from structural characteristics of the stimulus-shapes." *American Journal of Psychology* 76:589–597.

Francis, Clare Anne, and David R. Heise. 2008. "Emotions on the job: supporting and threatening face in work organizations." Pp. 107–129 In *Social Structure and Emotion*, edited by Jody Clay-Warner and Dawn Robinson. New York: Elsevier.

Friedkin, Noah E. 1998. *A Structural Theory of Social Influence*. New York: Cambridge University Press.

Garrett, Karen. 1982. "Child abuse: problems of definition." Pp. 177–204 in *Measuring Social Judgments: The Factorial Survey Approach*, edited by Peter H. Rossi and Steven L. Nock. Beverly Hills, CA: Sage Publications.

Gollob, Harry F. 1968. "Impression formation and word combination in sentences." *Journal of Personality and Social Psychology* 10:341–353.

———. 1974a. "Some tests of a social inference model." *Journal of Personality and Social Psychology* 29:157–172.

———. 1974b. "The subject–verb–object approach to social cognition." *Psychological Review* 81:286–321.

Gollob, Harry F., and Gregory W. Fischer. 1973. "Some relationships between social inference, cognitive balance, and change in impression." *Journal of Personality and Social Psychology* 26:16–22.

Gollob, Harry F., and B. B. Rossman. 1973. "Judgments of an actor's 'Power and ability to influence others'." *Journal of Experimental Social Psychology* 9:391–406.

Gollob, Harry F., Betty B. Rossman, and Robert P. Abelson. 1973. "Social inference as a function of the number of instances and consistency of information presented." *Journal of Personality and Social Psychology* 27:19–33.

Goodenough, Ward H. 1961. "Comments on cultural evolution." *Daedalus* 90:521–528.

Gordon, Robert A., James F. Short, Jr., Desmond S. Cartwright, and Fred. L. Strodtbeck. 1963. "Values and gang delinquency: a study of street corner groups." *American Journal of Sociology* 69:109–128.

Häder, Sabine, and Siegried Gabler. 2003. "Sampling and estimation." Pp. 117–134 in *Cross-Cultural Survey Methods*, edited by Janet A. Harkness, Fons J. R. van de Vijver, and Peter Mohler. New York: Wiley-Interscience.

Harkness, Janet A., Fons J. R. van de Vijver, and Peter Mohler. 2002. *Cross-Cultural Survey Methods*. New York: Wiley-Interscience.

Heise, David R. 1965. "Semantic differential profiles for 1,000 most frequent English words." *Psychological Monographs* 79: Issue 601.

———. 1966. "Sensitization of verbal response-dispositions by *n* Affiliation and *n* Achievement." *Journal of Verbal Learning and Verbal Behavior* 5:522–525.

———. 1969a. "Affective dynamics in simple sentences." *Journal of Personality and Social Psychology* 11:204–213.

———. 1969b. "Separating reliability and stability in test–retest correlation." *American Sociological Review* 34:93–101.

———. 1969c. "Some methodological issues in semantic differential research." *Psychological Bulletin* 72:406–422.

———. 1970a. "Potency dynamics in simple sentences." *Journal of Personality and Social Psychology* 16:48–54.

———. 1970b. "The semantic differential and attitude research." Pp. 235–253 in *Attitude Measurement*, edited by Gene Summers. Chicago: Rand McNally.

———. 1972. "Employing nominal variables, induced variables, and block variables in path analysis." *Sociological Methods and Research* 1:147–173.

———. 1974. "Some issues in sociological measurement." Pp. 1–16 in *Sociological Methodology: 1973–75*, edited by Herbert Costner. San Francisco: Jossey-Bass.

———. 1975. *Causal Analysis*. New York: Wiley.

———. 1977. "Social action as the control of affect." *Behavioral Science* 22:163–177.

———. 1978. *Computer-Assisted Analysis of Social Action: Use of Program INTERACT and SURVEY.UNC75*. Chapel Hill, NC: Institute for Research in the Social Sciences.

———. 1979. *Understanding Events: Affect and the Construction of Social Action*. New York: Cambridge University Press.

———. 1982a. "Face synthesizer." *Micro: The 6502/6809 Journal* 49:31–37.

———. 1982b. "Measuring attitudes with a PET." *BYTE: The Small Systems Journal* 7:208–246.

———. 1985. "Affect control theory: respecification, estimation, and tests of the formal model." *Journal of Mathematical Sociology* 11:191–222.

———. 1991. *OLS Equation Estimations for Interact*. Bloomington, IN: Department of Sociology, Indiana University.

———. 1997. Interact On-Line (Java applet). www.indiana.edu/~socpsy/ACT/interact/JavaInteract.html

———. 1998. "Conditions for empathic solidarity." Pp. 197–211 in *The Problem of Solidarity: Theories and Models*, edited by Patrick Doreian and Thomas J. Fararo. Amsterdam: Gordon and Breach.

———. 2001a. "Project Magellan: Collecting cross-cultural affective meanings via the Internet." *Electronic Journal of Sociology* 5.

———. 2001b. "Social measurement, classification and scaling." Pp. 13504–13508 in *International Encyclopedia of the Social and Behavioral Science*, edited by N. J. Smelser and Paul B. Baltes. New York: Pergamon.

———. 2004. "Enculturating agents with expressive role behavior." Pp. 127–142 in *Agent Culture: Human–Agent Interaction in a Multicultural World*, edited by Sabine Payer and Robert Trappl. Mahwah, NJ: Lawrence Erlbaum Associates.

———. 2005. *Magellan.Surveyor Documentation*. Bloomington, IN: Indiana University. www.indiana.edu/~socpsy/ACT/SurveyorDocumentation.htm

———. 2006. "Sentiment formation in social interaction." Pp. 189–211 in *Purpose, Meaning, and Action: Control Systems Theories in Sociology*, edited by Kent A. McClelland and Thomas J. Fararo. New York: Palgrave Macmillan.

———. 2007. *Expressive Order: Confirming Sentiments in Social Actions*. New York: Springer.

Heise, David R., and George W. Bohrnstedt. 1970. "Validity, invalidity, and reliability." Pp. 104–129 in *Sociological Methodology: 1970*, edited by Edgar Borgatta and George W. Bohrnstedt. San Francisco: Jossey-Bass.

Heise, David R., and Steven J. Lerner. 2006. "Affect control in international interactions." *Social Forces* 85:993–1010.

Heise, David R., and Neil J. MacKinnon. 1987. "Affective bases of likelihood perception." *Journal of Mathematical Sociology* 13:133–151.

Heise, David R., and Lynn Smith-Lovin. 1981. "Impressions of goodness, powerfulness and liveliness from discerned social events." *Social Psychology Quarterly* 44:93–106.

Heise, David R., and Lisa Thomas. 1989. "Predicting impressions created by combinations of emotion and social identity." *Social Psychology Quarterly* 52:141–148.

Herman-Kinney, Nancy J., and Joseph M. Verschaeve. 2003. "Methods of symbolic interactionism." Pp. 213–252 in *Handbook of Symbolic Interactionism*, edited by Larry T. Reynolds and Nancy J. Herman-Kinney. New York: Alta Mira Press, Rowman & Littlefield.

Holland, Dorothy. 1987. "Culture sharing across gender lines." *American Behavioral Scientist* 31:234–249.

Jenkins, J. J., W. A. Russell, and George C. Suci. 1958. "An atlas of semantic profiles for 360 words." *American Journal of Psychology* 71:688–699.

Kahneman, Daniel. 1963. "The semantic differential and the structure of inferences among attributes." *American Journal of Psychology* 76:554–567.

Kahneman, Daniel, and Dale T. Miller. 1986. "Norm theory: comparing reality to its alternatives." *Psychological Review* 93:136–153.

Kahneman, Daniel, Paul Slovic, and Amos Tversky (Eds.). 1982. *Judgment Under Uncertainty: Heuristics and Biases*. New York: Cambridge University Press.

Karabatsos, George, and William H. Batchelder. 2003. "Markov chain estimation for test theory without an answer key." *Psychometrika* 68:373–389.

Kashima, Yoshihisa. 2002. "Culture and self: a cultural dynamical analysis." Pp. 207–226 in *Self and Identity: Personal, Social, and Symbolic*, edited by Yoshihisa Kashima, Margaret Foddy, and Michael J. Platow. Mahwah NJ: Lawrence Erlbaum Associates.

Kish, Leslie. 1965. *Survey Sampling*. New York: Wiley.

Kiss, George R., Christine Armstrong, Robert Milroy, and J. R. I. Piper. 1973. "An associative thesaurus of English and its computer analysis." In *The Computer and Literary Studies*, edited by A. J. Aitken, R. W. Bailey, and N. Hamilton-Smith. Edinburgh, UK: University Press.

Kline, Rex B. 1998. *Principles and Practice of Structural Equation Modeling*. New York: Guilford Press.

Kroska, Amy, and Sarah K. Harkness. 2006. "Stigma sentiments and self-meanings: Exploring the modified labeling theory of mental illness." *Social Psychology Quarterly* 69:325–348.

Krosnick, Jon A., and Duane F. Alwin. 1987. "An evaluation of a cognitive theory of response-order effects in survey measurement." *Public Opinion Quarterly* 51:201–219.

Labov, William. 1972. *Language in the Inner City*. Philadelphia: University of Pennsylvania Press.

Labov, William, S. Ash, and C. Boberg. 2006. *The Atlas of North American English: Phonetics, Phonology and Sound Change*. Berlin: Mouton/de Gruyter.

Landau, Sidney I. (Ed.). 1975. *The Doubleday Dictionary*. Garden City, NY: Doubleday.

———. 2001. *Dictionaries: The Art and Craft of Lexicography*. New York: Cambridge University Press.

Landis, Daniel, P. McGrew, H. Day, J. Savage, and Tulsi Saral. 1976. "Word meanings in black and white." Pp. 45–80 in *Variations in Black and White Perceptions of the Social Environment*, edited by Harry C. Triandis. Urbana, IL: University of Illinois Press.

Landis, Daniel, and Tulsi Saral. 1978. "Atlas of American Black English" (computer printout). Urbana, IL: University of Illinois.

Lawson, Edwin D. 1973. "Men's first names, nicknames, and short names: a semantic differential analysis." *Names* 21:22–27.

Lawson, Edwin D., and Lynn M. Roeder. 1986. "Women's full names, short names and affectionate names: a semantic differential analysis." *Names* 34:175–184.

Lazarsfeld, Paul F. 1972. "The obligations of the 1950 pollster to the 1984 historian." Pp. 278–299 in *Qualitative Analysis: Historical and Critical Essays*, edited by Paul F. Lazarsfeld. Boston: Allyn and Bacon.

Liker, Jeffrey K. 1982. "Family prestige judgments: bringing in real-world complexities." Pp. 119–144 in *Measuring Social Judgments: The Factorial Survey Approach*, edited by Peter H. Rossi and Steven L. Nock. Beverly Hills, CA: Sage Publications.

Liu, James H., and Bibb Latané. 1998. "The catastrophic link between the importance and extremity of political attitudes." *Political Behavior* 20:105–126.

Lord, Frederic M., and Melvin R. Novick. 1968. *Statistical Theories of Mental Test Scores*. Reading, MA: Addison-Wesley.

MacKinnon, Neil J. 1985. *Affective Dynamics and Role Analysis* (report for SSHRC Project 410-81-0089). Guelph, Ontario, Canada: Department of Sociology and Anthropology, University of Guelph.

———. 1988. *The Attribution of Traits, Status Characteristics, and Emotions in Social Interaction* (report for SSHRC Project 410-81-0089). Guelph, Ontario, Canada: Department of Sociology and Anthropology, University of Guelph.

———. 1994. *Symbolic Interactionism as Affect Control*. Albany, NY: State University of New York Press.

MacKinnon, Neil J., and David R. Heise. 2010. *Self, Identity and Social Institutions*. New York: Palgrave.

MacKinnon, Neil J., and A. Luke. 2002. "Changes in identity attitudes as reflections of social and cultural change." *Canadian Journal of Sociology* 27:299–338.

Malone, Martin J. 2004. "Structure and affect: the influence of social structure on affective meaning in American kinship." *Social Psychology Quarterly* 67:203–216.

Mazzarella, William. 2004. "Culture, globalization, mediation." *Annual Review of Anthropology* 33:345–367.

McCulloch, A. Scott, and Peter A. Raynolds. 2007. *The Projective Differential*. projectivedifferential.com

McDougall, William. 1908. *An Introduction to Social Psychology*. London: Methuen.

Meudell, M. Bonner. 1982. "Household social standing: Dynamic and static dimensions." Pp. 69–94 in *Measuring Social Judgments: The Factorial Survey Approach*, edited by Peter H. Rossi and Steven L. Nock. Beverly Hills, CA: Sage Publications.

Miron, M. S. 1961. "The influence of instruction modification upon test–retest reliabilities of the semantic differential." *Educational and Psychological Measurement* 21:883–893.

Murdock, George Peter. 1949. *Social Structure*. New York: Macmillan.

Nock, Steven L. 1982. "Family social status: Consensus on characteristics." Pp. 95–118 in *Measuring Social Judgments: The Factorial Survey Approach*, edited by Peter H. Rossi and Steven L. Nock. Beverly Hills, CA: Sage Publications.

Norušis, Marija J. 2003. *SPSS 12.0 Statistical Procedures Companion*. Upper Saddle River, NJ: Prentice Hall.

Nunnally, Jum C. 1967. *Psychometric Theory*. New York: McGraw-Hill.

O'Brien, Lawrence J., Peter H. Rossi, and Richard C. Tessler. 1982. "How much is too much? Popular definitions of alcohol abuse." Pp. 235–252 in *Measuring Social Judgments: The Factorial Survey Approach*, edited by Peter H. Rossi and Steven L. Nock. Beverly Hills, CA: Sage Publications.

O'Rourke, Diane, Seymour Sudman, and Marya Ryan. 1996. "The growth of academic and not-for-profit survey research organizations." *Survey Research* 27:1–5.

Ortony, Andrew, and Gerald L. Clore. 1981. "Disentangling the affective lexicon." In *Third Annual Conference of the Cognitive Science Society*, Berkeley, CA.

Ortony, Andrew, Gerald L. Clore, and Mark Foss. 1987. "The referential structure of the affective lexicon." *Cognitive Science* 11:341–364.

Osgood, Charles E. 1953. *Method and Theory in Experimental Psychology*. New York: Oxford University Press.

———. 1962. "Studies on the generality of affective meaning systems." *American Psychologist* 17:10–28.

———. 1974. "Probing subjective culture: Part 2. Cross-cultural tool-using." *Journal of Communication* 24(2):235–269.

Osgood, Charles E., W. H. May, and M. S. Miron. 1975. *Cross-Cultural Universals of Affective Meaning*. Urbana, IL: University of Illinois Press.

Osgood, Charles E., George C. Suci, and Percy H. Tannenbaum. 1957. *The Measurement of Meaning*. Urbana, IL: University of Illinois Press.

Ragin, Charles C. 1987. *The Comparative Method. Moving Beyond Qualitative and Quantitative Strategies*. Berkeley, CA: University of California Press.

Rashotte, Lisa Slattery. 2001. "Some effects of demeanor on the meaning of behaviors in context." *Current Research in Social Psychology* 6.

———. 2002. "What does that smile mean? The meaning of nonverbal behaviors in social interaction." *Social Psychology Quarterly* 65:92–102.

———. 2003. "Written versus visual stimuli in the study of impression formation." *Social Science Research* 32:278–293.

Raynolds, Peter A., and Gennie H. Raynolds. 1989. "The 'JOG your right brain exercise' at ISAGA 88." In *Simulation–Gaming: On the Improvement of Competence in Dealing with Complexity, Uncertainty, and Value Conflicts*, edited by Jan H. Klabberg, Willem J. Scheper, Cees A. Takkenberg, and David Crookall. New York: Pergamon Press.

Raynolds, Peter A., Shiori Sakamoto, and Gennie H. Raynolds. 1988. "Consistent projective differential responses by American and Japanese students." *Perceptual and Motor Skills* 66:395–402.

Raynolds, Peter A., Shiori Sakamoto, and Robert Saxe. 1981. "Consistent responses by groups of subjects to projective differential items." *Perceptual and Motor Skills* 53:635–644.

Robinson, Dawn T., Lynn Smith-Lovin, and Olga Tsoudis. 1994. "Heinous crime or unfortunate accident? The effects of remorse on responses to mock criminal confessions." *Social Forces* 73:175–190.

Romney, A. Kimball. 1994. "When does consensus indicate cultural knowledge?" *CogSci News* 7:3–7.

———. 1999. "Culture consensus as a statistical model." *Current Anthropology* 40, Supplement:S103–S115.

Romney, A. Kimball, William H. Batchelder, and S. C Weller. 1987. "Recent applications of cultural consensus theory." *American Behavioral Science* 31:163–177.

Romney, A. Kimball, Carmella C. Moore, William H. Batchelder, and Ti-Lien Hsia. 2000. "Statistical methods for characterizing similarities and differences between semantic structures." *Proceedings of the National Academy of Sciences* 97:518–523.

Romney, A. Kimball, and Susan C. Weller. 1984. "Predicting informant accuracy from patterns of recall among inidividuals." *Social Networks* 6:59–77.

Romney, A. Kimball, S. C. Weller, and William H. Batchelder. 1986. "Culture as consensus: a theory of culture and informant accuracy." *American Anthropologist* 88:313–338.

Rossi, Alice S., and Peter H. Rossi. 1990. *Of Human Bonding: Parent–Child Relations Across the Life Course*. New York: Aldine de Gruyter.

Rossi, Peter H., and Andy B. Anderson. 1982. "The factorial survey approach: an introduction." Pp. 15–67 in *Measuring Social Judgments: The Factorial Survey Approach*, edited by Peter H. Rossi and Steven L. Nock. Beverly Hills, CA: Sage Publications.

Rossi, Peter H., and Richard A. Berk. 1985. "Varieties of normative consensus." *American Sociological Review* 50:333–347.

Rossi, Peter H., and Steven L. Nock (Eds.). 1982. *Measuring Social Judgments: The Factorial Survey Approach*. Beverly Hills, CA: Sage Publications.

Rummel, R. J. 1970. *Applied Factor Analysis*. Evanston, IL: Northwestern University Press.

Sanbonmatsu, David M., and Russell H. Fazio. 1990. "The role of attitudes in memory-based decision making." *Journal of Personality and Social Psychology* 59:614–622.

Schneider, Andreas. 2002. "Computer simulation of behavior prescriptions in multicultural corporations." *Organization Studies* 23:105–131.

Schneider, Andreas, and David R. Heise. 1995. "Simulating symbolic interaction." *Journal of Mathematical Sociology* 20:271–287.

Scholl, Wolfgang. 2008. *The Socio-emotional Basis of Human Cognition, Communication, and Interaction*. Berlin: Humboldt-University, Institute of Psychology.

Schonlau, Matthias, Ronald D. Fricker, Jr., and Marc N. Elliot. 2002. *Conducting Research Surveys via E-Mail and the Web*. Santa Monica, CA: Rand Corporation.

Schröder, Tobias, and Wolfgang Scholl. 2009. "Affective dynamics of leadership: An experimental test of affect control theory." *Social Psychology Quarterly* 72:180–197.

Sewell, Abigail A., and David R. Heise. 2009. "Racial differences in sentiments: Exploring variant cultures" Bloomington, IN: Department of Sociology, Indiana University.

Shepelak, Norma J., and Duane F. Alwin. 1986. "Beliefs about inequality and perceptions of distributive justice." *American Journal of Sociology* 51:30–46.

Simpson, Patricia A., and Linda K. Stroh. 2004. "Gender differences: emotional expression and feelings of personal inauthenticity." *Journal of Applied Psychology* 89:715–721.

Smith, Herman W. 1995. "Predicting stress in American–Japanese business relations." *Journal of Asian Business* 12:79–89.

———. 2002. "The dynamics of Japanese and American interpersonal events: behavioral settings versus personality traits." *Journal of Mathematical Sociology* 26:71–92.

Smith, Herman W., and Linda E. Francis. 2005. "Social versus self-directed events among Japanese and Americans: self-actualization, emotions, moods, and trait disposition labeling." *Social Forces* 84:821–830.

Smith, Herman W., Shuuichirou Ike, and Li Yeng. 2002. "Project Magellan redux: problems and solutions with collecting cross-cultural affective meanings via the Internet." *Electronic Journal of Sociology* 6.

Smith, Herman W., Takanori Matsuno, and Shuuichirou Ike. 2001. "The affective basis of attributional processes among Japanese and Americans." *Social Psychology Quarterly* 64:180–194.

Smith, Herman W., Takanori Matsuno, and Michio Umino. 1994. "How similar are impression-formation processes among Japanese and Americans?" *Social Psychology Quarterly* 57:124–139.

Smith-Lovin, Lynn. 1978. "Behavior settings and reactions to social scenarios: the impact of settings on the affective dynamics of interpersonal events" (Ph.D. dissentation). Chapel Hill, NC: University of North Carolina.

———. 1979. "Behavior settings and impressions formed from social scenarios." *Social Psychology Quarterly* 42:31–43.

———. 1987a. "The affective control of events within settings." *Journal of Mathematical Sociology* 13:71–101.

———. 1987b. "Impressions from events." *Journal of Mathematical Sociology* 13:35–70.

Smith-Lovin, Lynn, and David R. Heise. 1982. "A structural equation model of impression formation." Pp. 195–222 in *Multivariate Applications in the Social Sciences*, edited by Nancy Hirschberg and L. G. Humphreys. Hillsdale, NJ: Lawrence Erlbaum Associates.

——— (Eds.). 1988. *Analyzing Social Interaction: Advances in Affect Control Theory*. New York: Gordon and Breach.

Snider, James G., and Charles E. Osgood. 1969. *Semantic Differential Technique: A Sourcebook*. Chicago: Aldine.

Thomas, Lisa, and David R. Heise. 1995. "Mining error variance and hitting pay-dirt: discovering systematic variation in social sentiments." *Sociological Quarterly* 36:425–439.

Thomsen, Cynthia J., Eugene Borgida, and Howard Lavine. 1995. "The causes and consequences of personal involvement." Pp. 191–214 in *Attitude Strength: Antecedents and Consequences*, edited by Richard E. Petty and Jon A. Krosnick. Mahwah, NJ: Lawrence Erlbaum Associates.

Thurstone, L. L. 1928. "Attitudes can be measured." *American Journal of Sociology* 33:529–554.

Tourangeau, Roger, Lance Rips, and Kenneth A. Rasinski. 2000. *The Psychology of Survey Response*. New York: Cambridge University Press.

Tsoudis, Olga. 2000a. "The likelihood of victim restitution in mock cases: Are the 'rules of the game' different from prison and probation?" *Social Behavior and Personality* 28:483–500.

———. 2000b. "Relation of affect control theory to the sentencing of criminals." *Journal of Social Psychology* 140:473–485.

Tsoudis, Olga, and Lynn Smith-Lovin. 1998. "How bad was it? The effects of victim and perpetrator emotion on responses to criminal court vignettes." *Social Forces* 77:695–722.

———. 2001. "Criminal identity: the key to situational construals in mock criminal court cases." *Sociological Spectrum* 21:3–31.

Van de Geer, John P. 1971. *Introduction to Multivariate Analysis for the Social Sciences*. San Francisco: W.H. Freeman.

———. 1993. *Multivariate Analysis of Categorical Data: Applications*. Thousand Oaks, CA: Sage Publications.

von Schneidemesser, Luanne 2008. "DARE Webpage." http://polyglot.lss.wisc.edu/dare/dare.html.

Weinreich, Uriel. 1958. "Travels through semantic space." *Word* 14:346–366.

Weisberg, Herbert F. 2005. *The Total Survey Error Approach: A Guide to the New Science of Survey Research*. Chicago: University of Chicago Press.

Weller, Susan C. 1987. "Shared knowledge, intracultural variation, and knowledge aggregation." *American Behavioral Scientist* 31:178–193.

Weller, Susan C., A. Kimball Romney, and D. P. Orr. 1987. "The myth of a sub-culture of corporal punishment." *Human Organization* 46:39–47.

Werner, Oswald, and Joann Fenton. 1970. "Method and theory in ethnoscience or ethnoepistemology." Pp. 537–578 in *A Handbook of Method in Cultural Anthropology*, edited by Raoul Naroll and Ronald Cohen. Garden City, NY: Natural History Press, Doubleday.

Werts, C. E., K. G. Jöreskog, and R. I. Linn. 1971. "Comment on 'The estimation of measurement error in panel data'." *American Sociological Review* 36:110–112.

Wiley, D. E., and J. A. Wiley. 1970. "The estimation of measurement error in panel data." *American Sociological Review* 35:112–117.

Winkler, Robert L., and William L. Hays. 1970. *Statistics: Probability, Inference, and Decision*. New York: Holt, Rinehart and Winston.

Wundt, Wilhelm. 1897. *Outlines of Psychology*. Leipzig, Germany: Wilhelm Engelmann.

Index

Surveying Cultures: Discovering Shared Conceptions and Sentiments, By David R. Heise

Printed and bound by CPI Group (UK) Ltd, Croydon, CR0 4YY
16/09/2026
14963415-0001